T0192260

SpringerBriefs in Molecular Science

Biobased Polymers

Series editor

Patrick Navard, CNRS/Mines ParisTech, Sophia Antipolis, France

Published under the auspices of EPNOE*Springerbriefs in Biobased polymers covers all aspects of biobased polymer science, from the basis of this field starting from the living species in which they are synthetized (such as genetics, agronomy, plant biology) to the many applications they are used in (such as food, feed, engineering, construction, health, …) through to isolation and characterization, biosynthesis, biodegradation, chemical modifications, physical, chemical, mechanical and structural characterizations or biomimetic applications. All biobased polymers in all application sectors are welcome, either those produced in living species (like polysaccharides, proteins, lignin, …) or those that are rebuilt by chemists as in the case of many bioplastics.

Under the editorship of Patrick Navard and a panel of experts, the series will include contributions from many of the world's most authoritative biobased polymer scientists and professionals. Readers will gain an understanding of how given biobased polymers are made and what they can be used for. They will also be able to widen their knowledge and find new opportunities due to the multidisciplinary contributions.

This series is aimed at advanced undergraduates, academic and industrial researchers and professionals studying or using biobased polymers. Each brief will bear a general introduction enabling any reader to understand its topic.

*EPNOE The European Polysaccharide Network of Excellence (www.epnoe.eu) is a research and education network connecting academic, research institutions and companies focusing on polysaccharides and polysaccharide-related research and business.

More information about this series at http://www.springer.com/series/15056

Qingbin Guo · Lianzhong Ai
Steve W. Cui

Methodology for Structural Analysis of Polysaccharides

 Springer

Qingbin Guo
State Key Laboratory of Food Nutrition
  and Safety
Tianjin University of Science & Technology
Tianjin, China

Steve W. Cui
Guelph Research and Development Centre
Agriculture and Agri-Food Canada
Guelph, ON, Canada

Lianzhong Ai
School of Medical Instruments
  and Food Engineering
University of Shanghai for Science
  and Technology
Shanghai, China

ISSN 2191-5407   ISSN 2191-5415   (electronic)
SpringerBriefs in Molecular Science
ISSN 2510-3407   ISSN 2510-3415   (electronic)
Biobased Polymers
ISBN 978-3-319-96369-3   ISBN 978-3-319-96370-9   (eBook)
https://doi.org/10.1007/978-3-319-96370-9

Library of Congress Control Number: 2018949330

This Springer imprint is published by the registered company Springer Nature Switzerland AG
The registered company address is: Gewerbestrasse 11, 6330 Cham, Switzerland

Contents

Abbreviations

[η]	Intrinsic viscosity
1D and 2D NMR	1-dimension and 2-dimension Nuclear magnetic resonance
60P	High molecular weight polysaccharides from seed of Artemisia sphaerocephala Krasch
A2	Second virial coefficient
AFG	Acidic flaxseed gum
AMAC	9-aminoacridone
ANTS	8-aminonaphthalene-1,3,6-trisulfonate
APTS	8-aminopyrene-1,3,6-trisulfonate,
ASKP	Polysaccharides extracted from seed of Artemisia sphaerocephala Krasch
COSY	Correlation spectroscopy
DE	Degree of esterification
DEAE-C	Diethylaminoethyl cellulose
DHB	2,5-Dihydroxybenzoic
DLS	Dynamic light scattering
DMSO	Dimethyl sulfoxide
DP	Degree of polymerization
DS	Degree of substitution
EDTA	Ethylenediaminetetraacetic acid
EGSP	American ginseng pectic polysaccharides hydrolysates by pectinase
EGSPP	EGSP ethanol precipitate
EGSPS	EGSP ethanol supernatant
EPS	Exopolysaccharides
ESI	Electrospray ionization
EtOH	Ethanol
EV	Elution volume

FAB	Fast atom bombardment
FID	Flame ionization detector
FK	Flaxseed kernel
FKDF	Flaxseed kernel dietary fibres
FK-EP	EDTA-extracted polysaccharides in flaxseed kernel
FK-KPI	1M KOH-extracted polysaccharides in flaxseed kernel
FK-KPII	4M KOH-extracted polysaccharides in flaxseed kernel
FK-WP	Water-extracted polysaccharides in flaxseed kernel
FTIR	Fourier-transform infrared spectroscopy
GC-MS	Gas chromatography–mass spectrometry
GOS	Galactooligosaccharides
GSP	Pectic polysaccharides from American ginseng
HMBC	Heteronuclear multiple bond correlation
HMQC	Heteronuclear multiple-quantum coherence
HOHAHA	Homonuclear Hartmann-Hahn spectroscopy
HPAEC	High-performance anion exchange chromatography
HPLC	High-performance liquid chromatography
HPSEC	High-performance size exclusion chromatography
HSQC	Heteronuclear single quantum coherence
KPI-EPF-G24h	Endo-1,4-β-D-glucanase hydrolysis of KPI-EPF for 24 h
KPI-EPF-H1.5h	0.1 M TFA hydrolysis of KPI-EPF for 1.5 h
LIF	Laser-induced fluorescence
MALDI	Matrix-assisted laser desorption/ionization
MALS	Multi-angle light scattering
Mn	Number average molecular weight
Mp	Peak molecular weight
Mw	Weight average molecular weight
MWCO	Molecular weight cut-off
Mz	Zeta average molecular weight
NFG	Neutral flaxseed gum
NOESY	Nuclear overhauser effect spectroscopy
O-Ac	CH_3-C(=O)-O-
O-Me	CH_3-O-
PAD	Pulsed amperometric detector
PDI	Polydispersity index
PMAA	Partially methylated alditol acetates
PMP	1-phenyl-3-methyl-2-pyrazolin-5-one
RALS	Right angle light scattering
R_g	Radius of gyration
R_h	Hydrodynamic radius
RI	Refractive index
SFG	Soluble flaxseed gum
SLS	Static light scattering

TFA	Trifluoroacetic acid
TFMS	Trifluoromethanesulfonic acid
THF	Tetrahydrofuran
TOCSY	Total correlation spectroscopy
VWD	Variable wavelength detector

Chapter 1
Strategies for Structural Characterization of Polysaccharides

1.1 Complexity of Polysaccharide Molecules

Polysaccharide molecules are constructed by long chains of monosaccharide units bound together via glycosidic linkages. Naturally occurring polysaccharides can be simply classified into four categories according to the source differences: plant polysaccharides, seaweed polysaccharides, animal polysaccharides and microbial polysaccharides (Table 1.1). Each category has its own specific structural features, e.g. most hemicelluloses as plant polysaccharides contain 6 monosaccharides (rhamnose, arabinose, galactose, glucose, xylose and mannose) while some derivatized monosaccharides, e.g. anhydrogalactose, can be found in some seaweeds polysaccharides. The molecular structure offers the most fundamental knowledge for understanding the functional, conformational and physiological properties of polysaccharides, which in turn facilitate their food and non-food applications. However, structural characterization of polysaccharides is a challenging task due to the molecular complexity in terms of monosaccharide composition, glycosidic bonds (linkage patterns), degree of branching/branching position, α- or β-configurations, functional groups, molecular weight and molecular weight distribution.

Table 1.1 Classification of polysaccharides based on the sources

Polysaccharides categories	Examples
Plant polysaccharides	Cellulose, hemicellulose, exudate gums
Seaweed polysaccharides	Alginate, carrageenan, agar
Microbial polysaccharides	Xanthan gum, gellan, pullulan
Animal polysaccharides	Hyaluronic acid, chitin, glycogen

© Crown 2018
Q. Guo et al., *Methodology for Structural Analysis of Polysaccharides*,
Biobased Polymers, https://doi.org/10.1007/978-3-319-96370-9_1

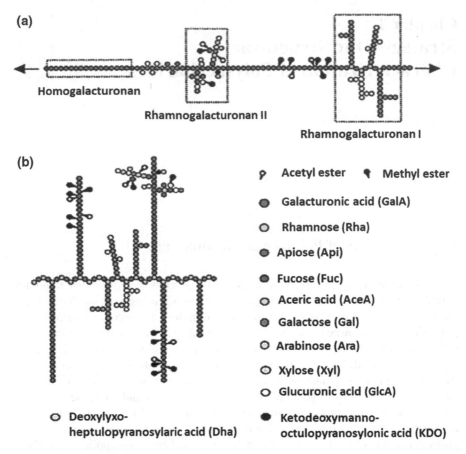

Fig. 1.1 Schematic chart of pectic polysaccharides. Adapted from Willats, Knox, and Mikkelsen (2006)

1.1.1 Monosaccharide Composition

Polysaccharide molecules can be constructed by either the same monosaccharide which termed as homopolysaccharide, or different monosaccharides/monosaccharide derivatives, designated heteropolysaccharide. For example, homopolysaccharides including cellulose, amylose and amylopectin are constructed by glucose only. Whereas most of the non-starch polysaccharides such as glucomannan (glucose and mannose), arabinoxylan (arabinose and xylose) and xyloglucan (xylose and glucose) are heteropolysaccharides. Taking pectin as an example, the types of monosaccharides/derivatives could be up to ten (Fig. 1.1).

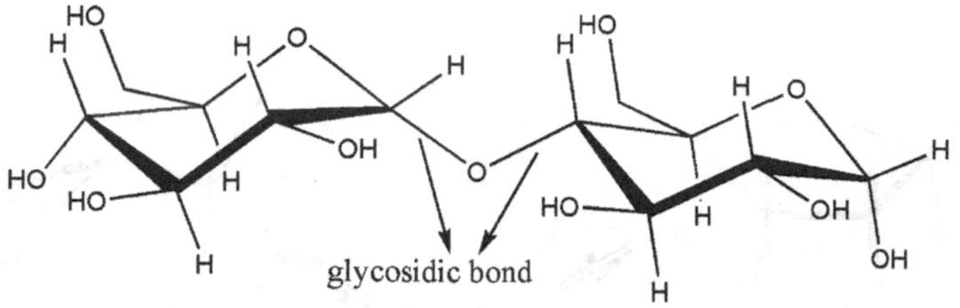

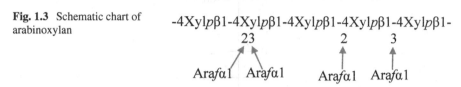

Fig. 1.2 Demonstration of $1 \rightarrow 4$ glycosidic bond from maltose

Fig. 1.3 Schematic chart of arabinoxylan

$$-4\text{Xyl}p\beta1\text{-}4\text{Xyl}p\beta1\text{-}4\text{Xyl}p\beta1\text{-}4\text{Xyl}p\beta1\text{-}4\text{Xyl}p\beta1\text{-}$$
$$\quad\quad\quad 23 \quad\quad\quad\quad\quad 2 \quad\quad\quad 3$$

$$\text{Ara}f\alpha1 \quad \text{Ara}f\alpha1 \quad\quad \text{Ara}f\alpha1 \quad \text{Ara}f\alpha1$$

1.1.2 Glycosidic Bonds

Varied glycosidic bonds (or referred as linkage patterns, Fig. 1.2) also contribute to the complexity of the polysaccharide molecules. For example, naturally occurring glucans contain $1 \rightarrow 4$ (β anomer in cellulose and α anomer in amylose), $1 \rightarrow 3$ & $1 \rightarrow 4$ (β anomer in cereal β-glucan), $1 \rightarrow 3$ & $1 \rightarrow 6$ (β anomer in curdlan) and $1 \rightarrow 4$ & $1 \rightarrow 6$ (α anomer in amylopectin and α anomer in pullulan) glycosidic bonds. The linkage pattern distributions varies among polysaccharides from different sources, which added more complexity to the polysaccharides molecules.

1.1.3 Branching Position and the Degree of Substitution

Polysaccharide molecules can be linear or branched. To fully understand the fine structure of branched polysaccharides, parameters include branching chain length, the degree of substitution and the branching distribution along the backbone need to be investigated. For example, the branching chain length distribution of amylopectin, which can be revealed by enzyme hydrolysis followed by HPAEC analysis (Chap. 4), are source-dependent; arabinoxylan molecules can be branched at O-2, O-3, or both O-2 and O-3 position of the xylan backbone as shown in Fig. 1.3. All these features increase the complexity for the structural characterization of branched polysaccharides.

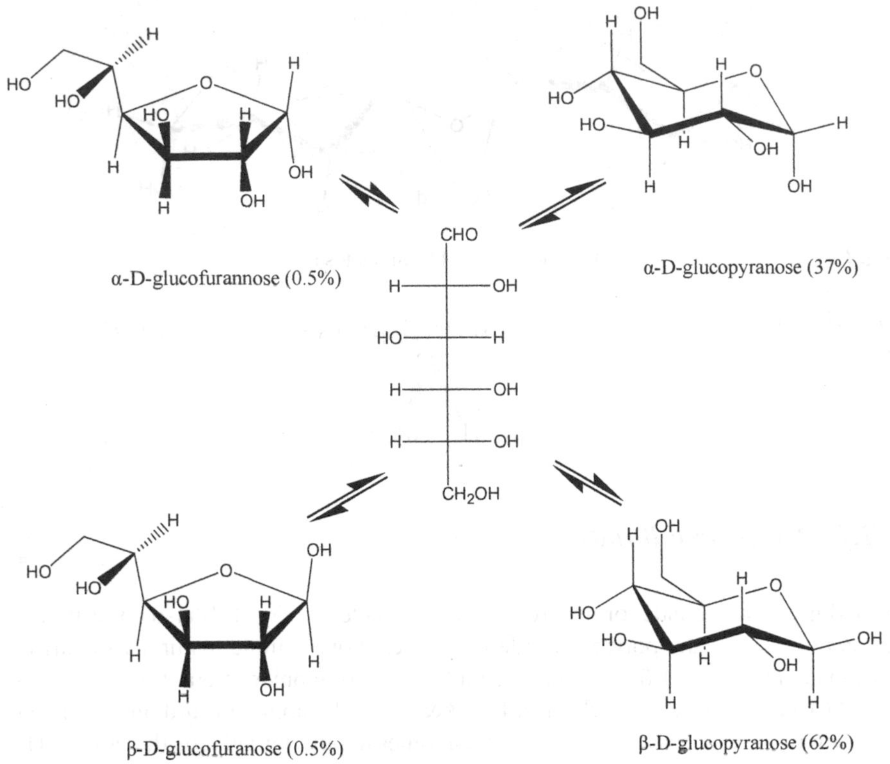

α-D-glucofurannose (0.5%)

α-D-glucopyranose (37%)

CHO
H————OH
HO————H
H————OH
H————OH
CH₂OH

β-D-glucofuranose (0.5%)

β-D-glucopyranose (62%)

Fig. 1.4 Configurations of free D-glucose in water solution

1.1.4 Configurations (α-Form and β-Form)

As monosaccharide, glucose adopts four different configurations in solution (Fig. 1.4). However, once it forms oligosaccharides/polysaccharides, the configurations are fixed.

The configuration differences could lead to totally different physicochemical properties of the polysaccharides; for example, cellulose with 1,4-β-D-GlcP as the repeating unit is water-insoluble; whereas amylose with 1,4-α-D-GlcP as the building block is readily dissolved in hot water. The configurations of sugar residues can be revealed by 1D NMR spectroscopy, which is covered in Chap. 7.

1.1.5 Special Functional Groups

The presence of special functional groups/monosaccharides derivatives in polysaccharide molecules highly increased the difficulties for their structural characterization, e.g. O-Me in pectin, O-Ac in galactomannan family, and N-Ac in some microbial polysaccharides (Ai et al., 2016). Similar to branched structure, characters such as abundance, position and distribution of these functional groups need to be investigated. These functional groups could highly influence the physicochemical properties of the polysaccharides. For instance, highly acetylated glucomannan from Dendrobium Officinale (traditional Chinese herbs) has been previously reported (Xing et al., 2013). This polysaccharide is readily dissolved in water. However, the water solubility is significantly decreased if the acetyl group is either decreased through alkaline treatment or increased through acetylation reaction. The immune modulating activity of this polysaccharide is also attributed to the acetyl group (Xing et al., 2013). Therefore, understanding the degree of substitution of these functional groups as well as their positions in polysaccharide molecules are critical in establishing their structure and function relationships.

1.1.6 Molecular Weight and Molecular Weight Distribution

Most of the naturally occurring polysaccharides exist in a distribution of chain lengths and molecular weights. The molecular weight of polysaccharides needs to be addressed statistically as average molecular weight calculated from the molecular weights of all the chains in the sample, e.g. weight average molecular weight (M_w), number average molecular weight (M_n). The molecular distribution can be expressed by the polydispersity index ($PDI = M_w/M_n$), which is relatively high for plant polysaccharides than that of microbial polysaccharides. The broader molecular weight distribution creates more difficulties for carrying out the structural characterization of polysaccharides (Cui & Wang, 2005).

In summary, the diversity and irregularity of molecular chains make the structural characterization of polysaccharides a most challenging task. This monograph provides some strategies for elucidating the fine structures of naturally occurring polysaccharides.

1.2 Strategies for Structural Characterization of Polysaccharides

Acquiring pure and narrow molecular weight dispersed polysaccharides samples are the prerequisites for their structural characterization. Therefore, extraction followed by purification and fractionation is always a must prior to structural characterization (Fig. 1.5). The molecular weight distribution of polysaccharides can be determined by HPSEC, according to which the polysaccharides with high PDI can be further fractionated to prepare more homogenous sample.

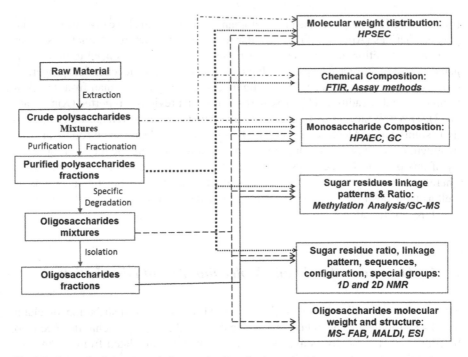

Fig. 1.5 Strategies for structural characterization of polysaccharides

The controlled molecular chain degradation such as enzymatical hydrolysis is normally required to convert large molecular weight polysaccharides into small fragments such as oligosaccharides; partial acidic hydrolysis could also produce oligosaccharides fragments. These fragments need to be further separated according to structural and molecular weight differences before carrying out structural analysis (Guo et al., 2011b). Monosaccharide composition analysis provides the first clue to the molecular structure. For example, high Mw polysaccharides from the seed of Artemisia sphaerocephala Krasch (ASKP) contained a high ratio of xylose, arabinose and glucuronic acid, which can be tentatively assigned to (glucurono)arabinoxylan family (Guo et al., 2011a). Methylation analysis combined with GC-MS uncovers the linkage patterns and molar ratios of the monomers while FT-IR can monitor and quantify the functional groups in the polysaccharide molecules. 1D &2D NMR spectroscopy, as a powerful technique, reveals structural features including linkage pattern, configuration, sequences as well as conformational properties of various sugar residues. In the following chapters, the structural characterizations of polysaccharides are presented in a systematic and step-wise manner.

References

Ai, L. Z., Guo, Q. B., Ding, H. H., Guo, B. H., Chen, W., & Cui, S. W. (2016). Structure characterization of exopolysaccharides from Lactobacillus casei LC2W from skim milk. *Food Hydrocolloids, 56*, 134–143.

Cui, S. W., & Wang, Q. (2005). Understanding the physical properties of food polysaccharides. In *Food carbohydrates: Chemistry, physical properties, and applications* (pp. 161–217). CRC Press.

Guo, Q., Cui, S. W., Wang, Q., Hu, X., Guo, Q., Kang, J., et al. (2011a). Extraction, fractionation and physicochemical characterization of water-soluble polysaccharides from Artemisia sphaerocephala Krasch seed. *Carbohydrate Polymers, 86*(2), 831–836.

Guo, Q., Cui, S. W., Wang, Q., Hu, X., Wu, Y., Kang, J., et al. (2011b). Structure characterization of high molecular weight heteropolysaccharide isolated from Artemisia sphaerocephala Krasch seed. *Carbohydrate Polymers, 86*(2), 742–746.

Willats, W. G. T., Knox, J. P., & Mikkelsen, J. D. (2006). Pectin: New insights into an old polymer are starting to gel. *Trends in Food Science & Technology, 17*(3), 97–104.

Xing, X., Cui, S. W., Nie, S., Phillips, G. O., Douglas Goff, H., & Wang, Q. (2013). A review of isolation process, structural characteristics, and bioactivities of water-soluble polysaccharides from Dendrobium plants. *Bioactive Carbohydrates and Dietary Fibre, 1*(2), 131–147.

Chapter 2
Polysaccharide Extraction and Fractionation

2.1 Extraction and Purification

Some polysaccharides can be collected directly from the ground endosperm of beans (guar gum) or husks (psyllium). Exudate gums such as gum arabic and gum ghatti can be picked up directly from the tree bark (Kang, Guo, Phillips, & Cui, 2014). However, most of the non-starch polysaccharides need to be extracted by water, mild base or acid solution (Guo et al., 2011, 2015). Assisted methods such as heating, microwave, and sonication were also extensively used to improve the extraction efficiency (Benko et al., 2007). Polysaccharides then can be recovered from the aqueous solution by either dialysis, ethanol precipitation, salt precipitation or directly freeze drying, accordingly.

Most of the cell wall polysaccharides can be sequentially extracted using water, chelating agent (EDTA) and alkaline solutions stepwise from weak to strong. For example, Ding et al. (2014) successfully extracted various types of hemicelluloses (arabinan, arabinoxylan, xyloglucan et al.) from de-oiled flaxseed kernel (Fig. 2.1). Results showed that most of the flaxseed dietary fibres were linked with other tissue constituents via noncovalent or covalent bonds and could not be directly extracted by hot water. After the treatment using chelating agents and alkaline solution, most of the fibres became water soluble (Ding et al., 2014).

For seaweed polysaccharides, e.g. carrageenan mild alkaline solutions are commonly used for the extraction. For example, the milled algal tissue needs to firstly mix with 0.1–0.2 M HCl to convert the alginate ions (Mg, Ca) into alginate acid, followed by alkaline (sodium carbonate or sodium hydroxide) soaking. The alginate acid then exists as sodium alginate, which is water-soluble and can be released into the solution. The sodium alginate can be recovered by either alcohol or calcium chloride precipitation as shown in Fig. 2.2. Carrageenans are a family of linear sulfated polysaccharides that are extracted from red edible seaweeds using alkaline water, e.g.

© Crown 2018   9
Q. Guo et al., *Methodology for Structural Analysis of Polysaccharides,*
Biobased Polymers, https://doi.org/10.1007/978-3-319-96370-9_2

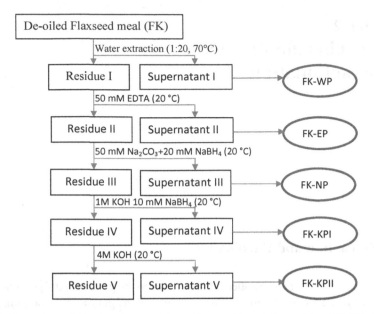

Fig. 2.1 Extraction and fractionation procedure of flaxseed kernel dietary fibres (FKDF). Adapted from Ding et al. (2014)

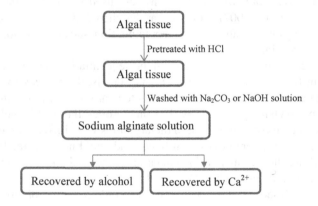

Fig. 2.2 Extraction of alginate

NaOH solution as solvents. However, the alkali selected may determine the particular type of carrageenan produced by the process which has important consequences for the properties of the resultant extract, in terms of dispersion, hydration, thickening and gelation (Cui, Izydorczyk, & Wang, 2005; Imeson, 2009).

Microbial polysaccharides can be classified into exopolysaccharides (EPS) and intracellular polysaccharides. Exopolysaccharides need to be isolated from culture media and microbial cell before analysis, e.g. cells and coagulate proteins can be removed by heating the culture medium for a short time (10 min) followed by centrifugation at room temperature. The leftover protein in the supernatant can be further removed by enzyme hydrolysis or trichloroacetic acid precipitation followed by centrifugation. EPS then can be obtained through dialysis followed by ethanol precipitation or freeze drying (Ai et al., 2008). For intracellular polysaccharides extraction, cells firstly need to be disrupted using homogenization, ultrasonication, or longer time boiling water treatment to release the polysaccharides (Fan et al., 2011; Zhang et al., 2017).

The purification process is normally required to remove contaminants such as coexisted starch and protein from the crude polysaccharides. Starch is removable by an enzymatic treatment with α-amylase and/or amyloglucosidase. Protein, if not covalently linked with polysaccharide molecule, can be excluded by either physical (seveg method (Navarini et al., 1999) and isoelectric precipitation), chemical (alkaline treatment) or enzymatic methods (protease hydrolysis). However, when protein is covalently linked with polysaccharide and is considered as part of the molecular structure, the chemical and enzymatical removal of proteins will not be considered. For example, the covalently linked protein in gum ghatti and gum arabic is the key to their excellent emulsification properties. Also, prior removal of lipid and lipid soluble material using ethanol or hexane soaking could greatly increase the water penetration and improve the extraction efficiency. Phenolic compounds are mostly enriched in the ethanol/hexane soaking solution.

2.2 Fractionation

Due to the nature of molecular weight distribution, polysaccharides need to be fractionated into narrower molecular weight fractions to better serve the structural characterization. In the meantime, the stepwise fractionation also could get rid of some contaminants, e.g. proteins. Differing in molecular size, charging properties, solubility, etc., polysaccharides can be separated by various fractionation techniques: ethanol sequential fractionation (Kang et al., 2011a), ammonium sulphate sequential precipitation (Guo et al., 2011; Li, Cui, & Wang, 2006; Wang, Wood, Huang, & Cui, 2003; Wu, Cui, Eskin, & Goff, 2009), membrane filtration (Guo et al., 2013) and column separation (Qian, Cui, Wu, & Goff, 2012).

2.2.1 Sequential Ammonium Sulphate Precipitation

Ammonium sulphate, which was originally used to precipitate protein, has been extensively applied for the fractionation of polysaccharides recently. Two mechanisms were proposed by Guo et al. (2011): (1) NH_4^+ is able to suppress the negative charges on the outer layer of polysaccharides which lead to polymer-polymer aggregation; (2) ammonium sulphate is able to reduce and/or change the solvent (water) properties which could lead to the precipitation of polysaccharide molecules.

As shown in Fig. 2.3, the fractionation of cereal β-glucan was carried out using sequential ammonium sulphate precipitation by Wang et al. (2003). Cereal β-glucan was successfully separated into six fractions of different molecular weight (F1–F6). The larger molecular weight molecules are easier to be precipitated by the relatively low concentration of salt. As no charged group exist, the precipitation only involves with the solubility difference of cereal β-glucan caused by the varied molecular weight. The ammonium sulphate precipitation method has also been applied for the acidic polysaccharides separation from Artemisia sphaerocephala Krasch seed (Guo et al., 2011).

2.2.2 Sequential Ethanol Precipitation

Ethanol is miscible with water. When adding it to polysaccharides solution, the interactions balance between polysaccharide-polysaccharide and polysaccharide-water molecules could be disrupted. This could induce aggregation of polysaccharides, and eventually precipitation. Molecules with relatively higher molecular weight and less branched structure tend to precipitate under the relatively lower concentration of ethanol while molecules with smaller molecular weight and highly branched structure need a higher concentration of ethanol to precipitate. It should be noted that the starting concentration and temperature of the polysaccharide solution play a critical role for the sequential ethanol precipitation, e.g. cold ethanol (4 °C) generally results in a better precipitation effect (Xing et al., 2015).

As shown in Fig. 2.4, ghatifolia (gum ghatti), one exudate gum, was successfully fractionated using sequential ethanol precipitation at room temperature by Kang et al. (2011a). Monosaccharide composition analysis was also conducted for these fractions as shown in Table 2.1. Results indicated that molecular structure of gum ghatti with a higher percentage of galactose and a lower percentage of arabinose was easier to be precipitated by ethanol. This was further confirmed in the following structural investigation conducted by Kang et al. (2011b, c): arabinose-based sugar residues were mostly located in the branching chain while galactose-based residues distributed mainly in the backbone. For the sequential fractionation purpose, ethanol should be slowly added to the solution to avoid the sudden increased local concentration.

2.2.3 Column Separation

Column separation has also been widely used for polysaccharides fractionation, for example, diethylaminoethyl cellulose (DEAE-C) or Sepharose column generally separate polysaccharides according to the distinction of their charged properties,

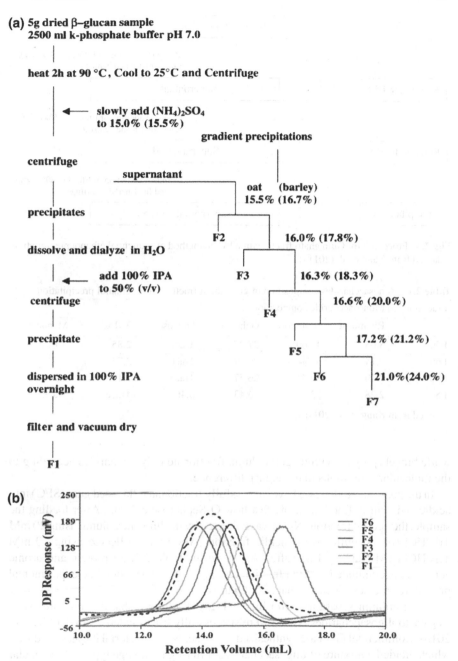

Fig. 2.3 Procedures of oat β-glucan fractionation (**a**) and the elution profiles from HPSEC (**b**). Adapted from Wang et al. (2003)

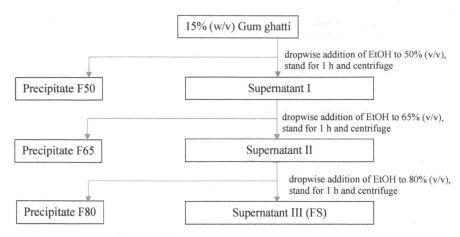

Fig. 2.4 Procedure for Gum ghatti fractionation by the method of sequential ethanol precipitation. Adapted from Kang et al. (2011a)

Table 2.1 Monosaccharides composition of gum ghatti fractions after ethanol precipitation

Fractions	Monosaccharide composition (wt%)					
	Rhamnose	Arabinose	Galactose	Glucose	Xylose	Mannose
F50	8.03	49.69	37.27	Trace	2.85	2.16
F65	2.14	63.44	29.09	Trace	3.74	1.60
F80	2.10	69.21	26.57	Trace	2.13	0.98
FS	2.28	72.85	16.43	8.46	Trace	Trace

Adapted from Kang et al. (2011a)

while biogel (polyacrylamide gel) column fractionate polysaccharides according to the molecular size (molecular weight) difference.

In the previous studies, we have successfully fractionated flaxseed gum (SFG) into acidic and neutral fractions using fast flow Q-Sepharose column. After loading the sample, the neutral fraction (NFG) was collected by flushing the column using 20 mM Tris/HCl (pH 8) buffer while acidic fraction (AFG) was collected using 20 mM Tris/HCl + 1M NaCl (pH 8) buffer. As shown in Table 2.2, all protein and uronic acid were accumulated in AFG fraction and not in NFG. The detailed experimental procedures are described in Chap. 10 (Qian et al., 2012).

Using column separation with biogel P-2 gel, oligosaccharides mixtures such as galactooligosaccharides (GOS) were successfully fractionated (Guo, Goff & Cui 2018). Commercial GOS are synthesized from lactose by bacterial β-galactosidases, which yielded a mixture of oligosaccharides differing in linkage type and molecular weight. The GOS mixtures need to be separated based on DP values before carrying out the structural analysis for each fraction. This can be achieved by Biogel P-2 Column (90 × 1 cm) fractionation as follows: 2 mL sample, diluted 2:1 with Milli-Q water, was loaded and eluted with water at a flow rate of 0.5 mL/min. The system

Table 2.2 Yield and chemical components of flaxseed gum and its fractions (%)

	SFG	NFG	AFG
Yield	9.7[a]	23.2[b]	27.3[b]
Protein	11.8	Nd	8.1
Uronic acid	23.0	1.8	38.7

Adapted from Qian et al. (2012)
SFG soluble flaxseed gum; *NFG and AFG* neutral and acidic fraction gum; *nd* not determined; all data were calculated on a dry basis
[a]Yield was based on flaxseed hull mass; [b]yield was based on SFG mass

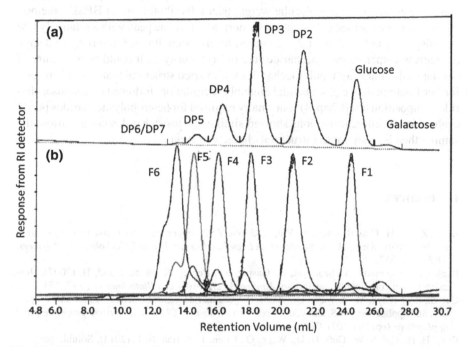

Fig. 2.5 Oligosaccharides profile of VITAGOS™ (a) and subfractions (b), obtained from HPSEC analysis coupled with refractive index detector. Adapted from Guo, Goff & Cui (2018)

was operated at 22 °C. Fractions were collected based on phenol-sulfuric assay (DuBois, Gilles, Hamilton, Rebers, & Smith, 1956) followed by HPSEC analysis using Rezex RSO-01 oligosaccharide Ag+ column coupled with RI detectors. The results are shown in Fig. 2.5. It is also worth noting that the biogel P gel beads have different size ranges, which could be selected for fractionation of polysaccharides (oligosaccharides) with different molecular weights, e.g. Biogel P-2, P-4, P-6, P-10, P-30 and P-100 can be used to fractionate polysaccharides with Mw range of 0.1–1.8, 0.8–4, 1–6, 1.5–20, 2.5–40, 3–60 and 5–100 kDa, respectively.

2.3 Purity Evaluation of Polysaccharide Samples Prior to Structural Analysis

As mentioned earlier, polysaccharides should be pure enough before structural analysis, the high purity refers to less contaminants and similar structural features. Various methods are involved in evaluating the purity of polysaccharides. Firstly, the total sugar content test (DuBois et al., 1956) can indicate whether the sample contains non-carbohydrate contaminates such as protein. Normally samples with high purity contain total sugar content above 90 wt%. Secondly, the purity of the polysaccharides can be determined by the molecular weight (size) distribution using HPSEC method (Chap. 3) as the purified polysaccharides normally has one peak with a relatively low polydispersity index (PDI < 1.2). However, till this step, it is still too risky to decide the narrow-dispersed polysaccharides are of high purity, as it could be a mixture of similar molecular weight polysaccharides with varied structural features. Therefore, further fractionation, e.g. sequential ethanol precipitation, followed by monosaccharide composition test (Chap. 4) is normally required for heteropolysaccharides purity evaluation. If different fractions showed similar monosaccharides composition, the sample then is ready for the structural characterization.

References

Ai, L., Zhang, H., Guo, B., Chen, W., Wu, Z., & Wu, Y. (2008). Preparation, partial characterization and bioactivity of exopolysaccharides from *Lactobacillus casei* LC2W. *Carbohydrate Polymers, 74*(3), 353–357.

Benko, Z., Andersson, A., Szengyel, Z., Gaspar, M., Reczey, K., & Stalbrand, H. (2007). Heat extraction of corn fiber hemicellulose. *Applied Biochemistry and Biotechnology, 137,* 253–265.

Cui, S., Izydorczyk, M., & Wang, Q. (2005). Polysaccharide gums: Structures, functional properties, and applications. In S. Cui (Ed.), *Food carbohydrates: Chemistry, physical properties, and applications* (pp. 263–307). CRC Press.

Ding, H. H., Cui, S. W., Goff, H. D., Wang, Q., Chen, J., & Han, N. F. (2014). Soluble polysaccharides from flaxseed kernel as a new source of dietary fibres: Extraction and physicochemical characterization. *Food Research International, 56,* 166–173.

DuBois, M., Gilles, K. A., Hamilton, J. K., Rebers, P. A., & Smith, F. (1956). Colorimetric method for determination of sugars and related substances. *Analytical Chemistry, 28*(3), 350–356.

Fan, K., Meng, G., Zhou, B., Deng, P., Liu, X., Jia, L., et al. (2011). Intracellular polysaccharide and its antioxidant activity by *Cordyceps militaris* SU-08. *Journal of Applied Polymer Science, 120*(3), 1744–1751.

Guo, Q., Goff, H. D., & Cui, S. W. (2018). Structural characterisation of galacto-oligosaccharides (VITAGOS™) sythesized by transgalactosylation of lactose. *Bioactive Carbohydrates and Dietary Fibre, 14,* 33–38.

Guo, Q., Cui, S. W., Kang, J., Ding, H. H., Wang, Q., & Wang, C. (2015). Non-starch polysaccharides from American ginseng: Physicochemical investigation and structural characterization. *Food Hydrocolloids, 44,* 320–327.

Guo, Q., Cui, S. W., Wang, Q., Hu, X., Guo, Q., Kang, J., et al. (2011). Extraction, fractionation and physicochemical characterization of water-soluble polysaccharides from Artemisia sphaerocephala Krasch seed. *Carbohydrate Polymers, 86*(2), 831–836.

Guo, Q., Wang, Q., Cui, S. W., Kang, J., Hu, X., Xing, X., et al. (2013). Conformational properties of high molecular weight heteropolysaccharide isolated from seeds of Artemisia sphaerocephala Krasch. *Food Hydrocolloids, 32*(1), 155–161.

Imeson, A. P. (2009). *7—Carrageenan and furcellaran. Handbook of hydrocolloids* (2nd ed., pp. 164–185). Woodhead Publishing.

Kang, J., Cui, S. W., Chen, J., Phillips, G. O., Wu, Y., & Wang, Q. (2011a). New studies on gum ghatti (*Anogeissus latifolia*) part I. Fractionation, chemical and physical characterization of the gum. *Food Hydrocolloids, 25*(8), 1984–1990.

Kang, J., Cui, S. W., Phillips, G. O., Chen, J., Guo, Q., & Wang, Q. (2011b). New studies on gum ghatti (*Anogeissus latifolia*) part II. Structure characterization of an arabinogalactan from the gum by 1D, 2D NMR spectroscopy and methylation analysis. *Food Hydrocolloids, 25*(8), 1991–1998.

Kang, J., Cui, S. W., Phillips, G. O., Chen, J., Guo, Q., & Wang, Q. (2011c). New studies on gum ghatti (*Anogeissus latifolia*) Part III: Structure characterization of a globular polysaccharide fraction by 1D, 2D NMR spectroscopy and methylation analysis. *Food Hydrocolloids, 25*(8), 1999–2007.

Kang, J., Guo, Q., Phillips, G. O., & Cui, S. W. (2014). Understanding the structure–emulsification relationship of gum ghatti—A review of recent advances. *Food Hydrocolloids, 42*, Part 1(0), 187–195.

Li, W., Cui, S. W., & Wang, Q. (2006). Solution and conformational properties of wheat β-D-Glucans studied by light scattering and viscometry. *Biomacromolecules, 7*(2), 446–452.

Navarini, L., Gilli, R., Gombac, V., Abatangelo, A., Bosco, M., & Toffanin, R. (1999). Polysaccharides from hot water extracts of roasted Coffea arabica beans: Isolation and characterization. *Carbohydrate Polymers, 40*(1), 71–81.

Qian, K. Y., Cui, S. W., Wu, Y., & Goff, H. D. (2012). Flaxseed gum from flaxseed hulls: Extraction, fractionation, and characterization. *Food Hydrocolloids, 28*(2), 275–283.

Wang, Q., Wood, P. J., Huang, X., & Cui, W. (2003). Preparation and characterization of molecular weight standards of low polydispersity from oat and barley (1-3)(1-4)-beta-glucan. *Food Hydrocolloids, 17*(6), 845–853.

Wu, Y., Cui, W., Eskin, N. A. M., & Goff, H. D. (2009). Fractionation and partial characterization of non-pectic polysaccharides from yellow mustard mucilage. *Food Hydrocolloids, 23*(6), 1535–1541.

Xing, X., Cui, S. W., Nie, S., Phillips, G. O., Goff, H. D., & Wang, Q. (2015). Study on Dendrobium officinale O-acetyl-glucomannan (Dendronan((R))): Part II. Fine structures of O-acetylated residues. *Carbohydrate Polymers, 117*, 422–433.

Zhang, C., Gao, Z., Hu, C., Zhang, J., Sun, X., Rong, C., et al. (2017). Antioxidant, antibacterial and anti-aging activities of intracellular zinc polysaccharides from Grifola frondosa SH-05. *International Journal of Biological Macromolecules, 95*, 778–787.

Chapter 3
Molecular Weight Distribution and Conformational Properties of Polysaccharides

As a fundamental characteristic, the molecular weight and molecular weight distribution of polysaccharides are important in relation to many physical properties. Average molecular weight and polydispersity index are required to quantify the molecular weight and its distribution. The equation of number average molecular weight (Mn), weight average molecular weight (Mw), zeta average molecular weight (Mz) and polydispersity index (PDI) are demonstrated as follows (Eqs. 3.1–3.4), whereas the C_i represents the concentration of molecules which has the molecular weight of M_i.

$$Mn = \frac{\sum C_i}{\sum \frac{C_i}{M_i}} \tag{3.1}$$

$$Mw = \frac{\sum M_i C_i}{\sum C_i} \tag{3.2}$$

$$Mz = \frac{\sum M_i^2 C_i}{\sum M_i C_i} \tag{3.3}$$

$$PDI = \frac{Mw}{Mn} \tag{3.4}$$

Different molecular weight determination methods such as membrane osmometry method, ultracentrifugation (sedimentation) have been well summarized by Cui and Wang (2005), and are not addressed in this chapter. The main focuses of this chapter are the triple-detector high-performance size exclusion chromatography (HPSEC) and batch mode multi-angle light scattering technique.

© Crown 2018

Q. Guo et al., *Methodology for Structural Analysis of Polysaccharides,*
Biobased Polymers, https://doi.org/10.1007/978-3-319-96370-9_3

3.1 Triple Detector HPSEC

3.1.1 Separation Principle of Size Exclusion Chromatography and Three Calibration Methods of Molecular Weight Determination

Size exclusion chromatography, also known as gel permeation chromatography, has been extensively applied for the fractionation of polymers based on molecular size (Guo & Chang, 2017; Yin et al., 2015). The column of SEC consists of a hollow tube tightly packed with extremely small porous polymer beads. When the polymer solution travels through the column, larger molecules elute much faster than smaller molecules as larger particles simply pass by the pores rather than going through the small pores in the beads.

Three calibration methods, including conventional, universal, and multi-detector-calibration methods, are generally used for the SEC methods, depending on the types and numbers of SEC detectors equipped. The conventional calibration only uses a concentration detector, e.g. refractive index (RI) and/or UV detector, and determines the molecular weight distributions based on the elution volumes/(EV)-molecular weight relation. Standards with different molecular weights are required and the obtained molecular weight need to be reported relative to the standards; this is because the conformation of the standards affects the calculation, meaning different types of standards could result in different calculated results. The universal calibration method is based on a unique relationship between the hydrodynamic volume of a macromolecule and its EV or retention time in the SEC columns (Grubisic, Rempp, & Benoit, 1967). Both viscometer and concentration detectors are required for the universal calibration method. The multi-detector-calibration method is the most frequently used method. By using an integrated RI, multi-angle light scattering (MALS), and four-capillary online viscometer, the molecular weight (Mn, Mw, Mz & Mp), intrinsic viscosity [η], polydispersity index (PDI) and the Mark-Houwink constant (α) of polysaccharides can be obtained (Guo et al., 2013). In addition, the aggregation effects in HPSEC measurement could be decreased to an fairly low level (Li et al., 2006), since the relatively low concentration (1 or 2 mg/mL), elevated temperature (40 °C) and the forced shear flow effect allows for the dissociation of the aggregation. Also, the aggregation region (if exist) could be easily excluded as we can freely select various peak regions for calculation.

3.1.2 Conformational Properties of Polysaccharides by HPSEC

The triple-detector HPSEC method has been widely applied for the understanding of conformational properties of polysaccharides (Guo et al., 2013; Yin et al., 2015).

As aforementioned, combining RI, RALS with an online viscometer, both molecular weight and intrinsic viscosity [η] can be obtained, based on which the molecular size and conformational properties of the polysaccharides can be determined.

Firstly, the double logarithmic plot of the molecular weight versus intrinsic viscosity can be well described using the Mark-Houwink Eq. (3.5)

$$[\eta] = kM_v^{\alpha} \tag{3.5}$$

where k and α are used to investigate the corresponding conformation of polysaccharides. The exponent α increased with the increasing chain stiffness: the α of 0.5–0.8 is generally adopted for a random coil conformation under aqueous solution, while polysaccharides with α value lower than 0.5 and above 0.8 are characteristics for spherical and rigid rod conformation, respectively. It should be noted that the α value is also affected by many other factors, such as solvent quality, and temperature. According to Morris and Ross-Murphy (1981), lower α value favors poor solvent and higher α indicating a good solvent.

As shown in Fig. 3.1a, the double logarithmic plot of the molecular weight versus intrinsic viscosity of polysaccharides (isolated from the seed of *Artemisia sphaerocephala* Krasch.) was not completely linear, i.e., a decreased slope was observed with increasing Mw, likely because of the increased percentage of side chains in the high Mw fraction. Hence, two linear regressions were used to fit the curve, the slope (α) were obtained as 0.82 and 0.32 in the log Mw range of 5.3–5.82 and 5.82–6.3 respectively. The intercept (k) was also detected (3.2). This indicated that polysaccharide molecules in aqueous solution exhibited random coil conformation in low Mw range and close to spherical conformation in high Mw range.

The hydrodynamic radius (R_h) can be obtained from HPSEC, based on the following equation (Eqs. 3.6–3.7).

$$R_h = \frac{([\eta]M)^{1/3}}{3.9} \tag{3.6}$$

The radius of gyration (R_g) is calculated from the Flory-Fox equation (Flory, 1953)

$$R_g = 1/\sqrt{6}\left(\frac{[\eta]M}{\Phi}\right)^{1/3} \tag{3.7}$$

where Φ is the Flory viscosity constant (roughly 2.6×10^{26} kg^{-1} for random coils).

The relationships of R_h versus M_w, R_g versus M_w and R_g/R_h versus M_w could also reflect the conformational properties of polysaccharides, which could be easily constructed using triple-detector HPSEC methods (Guo et al., 2013). As shown in Fig. 3.1b, using linear regression, the slopes of R_g versus M_w and R_h versus M_w were 0.49 and 0.55 respectively. This also confirmed that this polysaccharide in aqueous solution exhibited random coil conformation (Harding, Abdelhameed, & Morris,

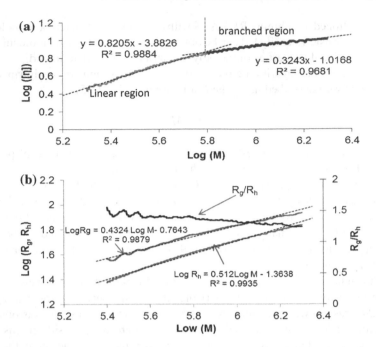

Fig. 3.1 Logarithmic plot of the molecular weight versus intrinsic viscosity (**a**), radius of gyration, hydrodynamic radius, the ratio of R_g/R_h (**b**) of polysaccharides (from the seed of *Artemisia sphaerocephala* Krasch.) in aqueous solution. Adapted from Guo et al., (2013)

2011). The plot of R_g/R_h versus Log M_w was also carried out and the value of ρ was in the range of 1.3–1.6 (Burchard, 1994), indicating a random coil conformation.

3.1.3   Branched Structure of Polysaccharides

The branching characteristic is one of the important molecular parameters which determines various physicochemical properties of polysaccharides such as solubility, viscosity, etc. Branched polysaccharides can be of varied structures, for example, it could be short chain branching or long chain branching; the branching position could be randomly or evenly distributed along the main chain or branching chain. In most branched polymers the distribution of the degree of branching coexists with the distribution of molecular weight and possibly with other characteristics such as chemical composition. Here the degree of branching is termed as "number of branch units in a macromolecule or number of arms in a starlike macromolecule". All the above-mentioned uncertainty brings extra challenges for the structural characterization of polysaccharides.

HPSEC coupled with triple detectors has been used to understand the branching character of polysaccharides in terms of degree of branching and branching distribution. The fundamental principle is "at a given molecular weight the molecular size decreases with increasing degree of branching". In other words, at the same molecular weight, branched polymers have lower hydrodynamic volume, higher density and therefore lower intrinsic viscosity (Podzimek, 2011; Scholte, 1983; Wang, Kharchenko, Migler, & Zhu, 2004).

Zimm and Stockmayer (1949) utilized mean square radius for the determination of branching ratio (g) at given molecular weight M.

$$g_M = \frac{R_{br}^2}{R_{lin}^2}$$ (3.8)

From which the branching ratio is equal to 1 for linear polymers and decreases with increasing extent of branching. For polymers with a high degree of branching, the ratio approaches 0.1, but never zero.

The branching ratio g reflects the degree of branching in a distinct way according to the structural difference. For example, for randomly branched polymers with tri- (g_3) or tetra-functional (g_4) branch units, the ratio g is given by the equations:

$$g_3 = [\left(1 + \frac{m}{7}\right)^{\frac{1}{2}} + \frac{4m}{9\pi}]^{-\frac{1}{2}}$$ (3.9)

$$g_4 = [\left(1 + \frac{m}{6}\right)^{\frac{1}{2}} + \frac{4m}{3\pi}]^{-\frac{1}{2}}$$ (3.10)

whereas m is the average number of branching per molecule.

An alternative branching ratio that determined by the intrinsic viscosity of the branched molecule over linear molecule under same molecular weight has also been extensively used (Eq. 3.11)

$$g_M' = \frac{[\eta]_{M,br}}{[\eta]_{M,lin}}$$ (3.11)

The relationship of g' and g is as follows: $g_M' = g^b$ whereas b refers as a structural factor, which is affected by parameters such as branching ratio, solvent, temperature and molecular weight.

The branching distribution of polysaccharides with molecular weight can be demonstrated by triple detector HPSEC through the relationships between intrinsic viscosity and molecular weight as shown in Fig. 3.1b. For the detailed information about polymer branching characterization, please refer to the book chapter by Podzimek (2011). The degree of branching of polysaccharides could also be determined by methylation analysis as shown in Chap. 6.

3.2 Batch Mode Multi-angle Light Scattering

Different from the triple-detector HPSEC, which only contains fixed angle detector, e.g. low angle (7°) and right angle (90°) light scattering detector, the multi-angle light scattering allows us to be able to set the range of the detection angles as the detector is attached with a rotation arm. Based on which both static (SLS) and dynamic light scattering (DLS) methods can be applied.

Static light scattering can be applied for the determination of weight average molecular weight M_w, radius of gyration R_g, and the second virial coefficient A_2 using the following equation:

$$Kc/R_\theta = 1/M_w + 1/3(R_g^2/M_w)q^2 + 2A_2c \qquad (3.12)$$

where K is an optical contrast factor, c is the polymer concentration, R_θ is the Rayleigh ratio (normalized scattering intensity), and the scattering vector q is defined as

$$q = 4\pi n_0 \sin(\theta/2)/\lambda_0 \qquad (3.13)$$

with n_0 is the refractive index and λ_0 is the wavelength in vacuum.

The static light scattering data is usually analyzed by Zimm plot approach, which is constructed by plotting Kc/R_θ versus q^2+kc, the M_w, R_g and A_2 can then be extracted via two extrapolations (Fig. 3.3). Extrapolating the angular measurements to zero angle for each concentration measurement at each straight line in c, the slope of this line yields second virial coefficient A_2. Extrapolating the concentration measurements to zero concentration for each angle results in another straight line, the slope divided by the intercept of this line yields R_g.

The dynamic light scattering technique is used mainly to extract the hydrodynamic radius (R_h) as well as the size distribution of the polymers. In dilute solutions, the intensity correlation function $G_2(t)$ can be related to the electric field correlation function $g_1(t)$ by the Siegert relationship:

$$G_2(t) = B(1 + f^2|g_1(t)|^2) \qquad (3.14)$$

where B is the baseline as determined by the infinite delay-time point, f, less or equal to one, determines the intercept-to-baseline ratio and depends on the quality of the instrument set up. For simple monodisperse particles, $g_1(t)$ decays as a single exponential with decay constant Γ

$$g_1(t) = \exp(-\Gamma t) \qquad (3.15)$$

$$\Gamma = Dq^2 \qquad (3.16)$$

with D representing the diffusion coefficient and q the magnitude of the scattering wave vector.

Fig. 3.2 The particle size distribution of high Mw polysaccharide from the seed of *Artemisia sphaerocephala* Krasch., determined by dynamic light scattering **a** in pure water, **b** in 0.09 M NaCl, **c** in 5 M Urea and **d** in 0.5 M NaOH. Adapted from Guo et al., (2013)

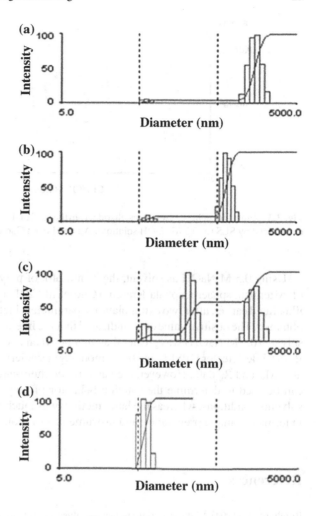

The hydrodynamic radius (R_h) can be calculated by the Stokes-Einstein equation:

$$D = kT/6\pi\eta R_h \tag{3.17}$$

where k is the Boltzmann's constant, T is the absolute temperature and η is the viscosity of the solvent.

It should be noted that polysaccharides easily form aggregates in aqueous solution, which greatly affect the light scattering results. Therefore, the elimination of aggregates need to be carried out prior to the test (Burchard, 1994; Li et al., 2006). As shown in Fig. 3.2, different types of aqueous solution including 0.09 M NaCl, 5 M urea and 0.5 M NaOH were compared for their aggregates elimination ability; only 0.5 M NaOH solution demonstrated monodispersed size distribution with the apparent size diameter (R_h) of 33.6 nm (Fig. 3.2).

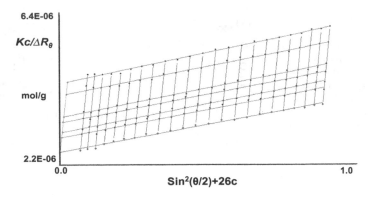

Fig. 3.3 Zimm plot of high Mw polysaccharide from the seed of *Artemisia sphaerocephala* Krasch., determined by SLS (in 0.5 M NaOH solution). Adapted from Guo et al., (2013)

Using 0.5 M NaOH as solvent, the Zimm plot of polysaccharides from the seed of Artemisia sphaerocephala Krasch. (Guo et al., 2013) are shown in Fig. 3.3. As illustrated earlier, after two extrapolations, parameters including M_w, R_g and A_2 were obtained. The positive value of A_2 indicated that the interaction between molecule and solvent (0.5 M NaOH) was greater than that between two polysaccharide molecules. Both triple detector HPSEC and batch mode light scattering methods could yield reliable M_w and R_g data. However, the batch mode multi-angle light scattering method can be used to determine the solution behavior of polysaccharides under different solvents conditions, whereas the latter method was much more convenient and faster in terms of sample preparation and instrument operation.

References

Burchard, W. (1994). Light scattering studies on polysaccharide solutions. *Papier, 48*(12), 755–764.

Cui, S. W., & Wang, Q. (2005). Understanding the physical properties of food polysaccharides. *Food carbohydrates: Chemistry, physical properties, and applications* (pp. 161–217). Boca Raton: CRC Press.

Flory, P. J. (1953). *Principles of polymer chemistry*. Ithaca, NY: Cornell University Press.

Grubisic, Z., Rempp, P., & Benoit, H. (1967). A universal calibration for gel permeation chromatography. *Journal of Polymer Science Part B: Polymer Letters, 5*(9), 753–759.

Guo, Q., & Chang, S. (2017). Tetra-detector size exclusion chromatography characterization of molecular and solution properties of soluble microbial polysaccharides from an anaerobic membrane bioreactor. *Frontiers of Environmental Science & Engineering, 11*(2), 16.

Guo, Q., Wang, Q., Cui, S. W., Kang, J., Hu, X., Xing, X., & Yada, R. Y. (2013). Conformational properties of high molecular weight heteropolysaccharide isolated from seeds of *Artemisia sphaerocephala* Krasch. *Food Hydrocolloids, 32*(1), 155–161.

Harding, S. E., Abdelhameed, A. S., & Morris, G. A. (2011). On the hydrodynamic analysis of conformation in mixed biopolymer systems. *Polymer International, 60*(1), 2–8.

Li, W., Wang, Q., Cui, S. W., Huang, X., & Kakuda, Y. (2006). Elimination of aggregates of (1-3) (1-4)-beta-D-glucan in dilute solutions for light scattering and size exclusion chromatography study. *Food Hydrocolloids, 20*(2–3), 361–368.

Morris, E. R., & Ross-Murphy, S. B. (1981). Chain flexibility of polysaccharides and glycoproteins from viscosity measurements. *Techniques in Carbohydrate Metabolism, B310,* 1–46.

Podzimek, S. (2011). Characterization of branched polymers. *Light scattering, size exclusion chromatography and asymmetric flow field flow fractionation* (pp. 307–345). Hoboken: Wiley.

Scholte, T. G. (1983). Characterization of long-chain branching in polymers. In J. V. Dawkins (Ed.), *Developments in polymer characterisation* (Vol. 4). UK: Applied Science, Barking.

Wang, W.-J., Kharchenko, S., Migler, K., & Zhu, S. (2004). Triple-detector GPC characterization and processing behavior of long-chain-branched polyethylene prepared by solution polymerization with constrained geometry catalyst. *Polymer, 45*(19), 6495–6505.

Yin, J. Y., Nie, S. P., Guo, Q. B., Wang, Q., Cui, S. W., & Xie, M. Y. (2015). Effect of calcium on solution and conformational characteristics of polysaccharide from seeds of *Plantago asiatica* L. *Carbohydrate Polymers, 124,* 331–336.

Zimm, B. H., & Stockmayer, W. H. (1949). The dimensions of chain molecules containing branches and rings. *Journal of Chemical Physics, 17,* 1301.

Chapter 4
Monosaccharide Composition Analysis

Monosaccharide composition analysis is critical for understanding the structure of polysaccharides, as it could provide the first clue of the structural information. For example, polysaccharides containing arabinose and xylose are mostly assigned to arabinoxylan family; polysaccharides composed of galactose and mannose mostly belong to galactomannan group; pectic polysaccharides are mostly constructed by galacturonic acid and some neutral monosaccharides. Monosaccharide composition analysis contains three steps: complete hydrolysis, separations and detections.

4.1 Complete Hydrolysis

Hydrolysis step is required to convert polysaccharides into monosaccharides before analysis. Sulfuric acid and trifluoroacetic acid (TFA) are commonly used for hydrolysis. For example, most of the neutral polysaccharides can be hydrolyzed by 1 M sulfuric acid at 100 °C for 2 h (Kang et al., 2011). TFA can be removed after hydrolysis via self-evaporation, therefore it does not interfere with the column separation and detection (Liu et al., 2014), however, the hydrolysis capability of TFA is relatively weak compared to sulfuric towards fibrous substrates such as wheat bran, straw, and microcrystalline cellulose (Brummer & Cui, 2005). Also, the stability of released monosaccharides varies which affects the hydrolysis conditions. For example, under the TFA hydrolysis, basic sugars (amino sugars, e.g. chitosan) were reluctant to be hydrolyzed, but the released monomers are stable; the acidic sugars (alginate) were easy to be hydrolyzed while the released monomers are unstable; for neutral sugar, the release and stability of the monomer were between basic sugar and acidic sugar (Wang, Zhao, Pu & Luan, 2016). Three types of hydrolysis reactions are recommended for different polysaccharides/glycoproteins according to Starr, Masada, Hague, Skop, and Klock (1996): using 0.1 M trifluoroacetic acid (TFA) at 80 °C for 30–60 min for sialic acids, and 2 M TFA/4–5 h for neutral-sugars, or 4 M HCl/3–5 h for amino-sugars, both at 100 °C.

© Crown 2018

Q. Guo et al., *Methodology for Structural Analysis of Polysaccharides*, Biobased Polymers, https://doi.org/10.1007/978-3-319-96370-9_4

In addition to acidic hydrolysis, enzymatic hydrolysis method has also been previously reported and demonstrated its advantages such as specific to linkages and not interfere with the detection due to the leftover ions such as sulphate ions from sulfuric acid. For example, the enzymes involved in the starch analysis including α-amylase, amyloglucosidase, pullulanase, and β-amylase. For non-starch polysaccharides, enzymes such as xylanase, cellulase (β-glucosidase) and pectinase are also widely used which target the linkages of β1→4xy1P, β1→4glc$P$ and α1→4gal$p$A, respectively. However, enzymes are mostly used to convert polysaccharides into oligomers (Ding et al., 2016), its use for monosaccharides production has been highly limited by the structural complexity in terms of branching and mixtures of various linkages pattern of the polysaccharide molecules. In addition, the types of available carbohydrate enzymes are also limited which compromises their applications in carbohydrate analysis.

4.2 Separations and Detections

Monosaccharides released from the hydrolysis can be identified and quantified by different techniques such as anion-exchange chromatography with pulsed amperometric detector (PAD), Gas chromatography (GC) with flame ionization detector (FID) for vaporized sugar derivatives, reverse phase HPLC with a variable wavelength detector (VWD) and capillary electrophoresis, each method is described in detail as follows:

4.2.1 Anion-Exchange Chromatography

High-performance anion-exchange chromatography with a pulsed amperometric detector (PAD) has been widely used for monosaccharide, sugar alcohol and oligosaccharides analysis (Guo, Cui, Wang, Hu, Kang et al., 2012; Guo, Cui, Wang, Hu, Wu et al., 2011). The columns used for these methods are commonly coated with an anion exchange resin. For example, the Dionex (Sunnyvale, CA) PA1 column, optimized for the separation of mono-, di-, oligo-, and low molecular weight polysaccharides, is composed of 10 μm nonporous beads covered in a quaternary amine anion exchange material. The detailed composition of PA1 and other CarboPAc columns are listed in Table 4.1. The elution buffer is normally alkaline (NaOH or KOH) or sodium acetate and can be programmed. The separation principle is as follows: Under alkaline conditions, the hydroxyl groups are ionized to oxyanions. Due to the different location and number of –OH group, the pKa of ionized monosaccharides are different, which results in different affinity for the oppositely charged stationary phase and the mobile phase. This method does not need derivatization of monosaccharides and provides reasonable separation resolution. However, it is easily interfered by anions such as sulphate ions or other contaminants in the mixture including protein and lipid.

Table 4.1 Summary of column composition

Column	Composition
CarboPAc PA1	10 mm diameter substrate (polystyrene 2% cross-linked with divinylbenzene) agglomerated with 500 nm MicroBead quaternary ammonium functionalized latex (5% cross-linked)
CarboPAc PA10	10 mm diameter substrate (ethylvinylbenzene 55% cross-linked with divinylbenzene) agglomerated with 460 nm MicroBead quaternary ammonium ion (5% cross-linked)
CarboPAc PA20	6.5 mm diameter substrate (ethylvinylbenzene 55% cross-linked with divinylbenzene) agglomerated with 130 nm MicroBead difunctional quaternary ammonium ion (5% cross-linked)
CarboPAc PA100	8.5 mm diameter ethylvinylbenzene/divinylbenzene substrate (55% cross-linked) agglomerated with 275 nm MicroBead quaternary amine functionalized latex (6% cross-linked)
CarboPAc PA200	5.5 mm diameter ethylvinylbenzene/divinylbenzene substrate (55% cross-linked) agglomerated with 43 nm MicroBead quaternary amine functionalized latex (6% cross-linked)
CarboPAc MA1	7.5 mm diameter vinylbenzyl chloride/divinylbenzene macroporous substrate fully functionalized with an alkyl quaternary ammonium group (15% cross-linked)

With classic CarboPAc PA1 column and PAD detector, monosaccharides and oligosaccharides profiles can be determined as shown in Fig. 4.1. The programmed elution buffer used are as follows:

(1) For monosaccharide composition analysis (Fig. 4.1a) e.g. the separation of rhamnose, arabinose, galactose, glucose, xylose and mannose; the elution procedure is programmed as follows: 0–7 min: 8 mM NaOH, 7–35 min: 100% water.

(2) For oligosaccharides and monosaccharides mixtures such as galactose, glucose, sucrose, fructose, raffinose, stachyose, verbascose separation (Fig. 4.1b), the programmed elution buffer is: 0–25 min: 10 mM NaOH, 25–50 min: 100% water.

(3) For debranched starch (Fig. 4.1c), the programmed elution procedure is: 0–5 min, 110 mM NaOH linear decreased to 90 mM, NaOAc increased from 200 mM to 300 mM; 5–25 min, NaOH decreased from 90 to 60 mM, NaOAc increased from 300 mM to 400 mM; 25–30 min, NaOH decreased from 60 mM to 55 mM, NaOAc increased from 400 mM to 475 mM; 30–35 min, NaOH decreased from 55 mM to 0 mM (pure water), NaOAc increased from 475 mM to 750 mM; 35–45 min, running with 750 mM NaOAc.

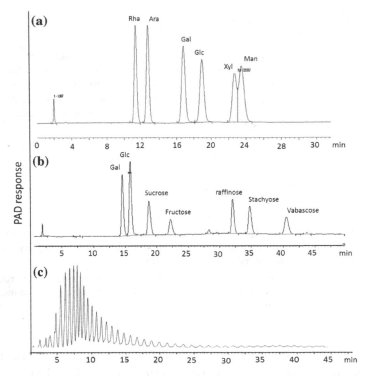

Fig. 4.1 HPAEC-PAD chromatogram showing separation of rhamnose, arabinose, glucose, galactose, xylose, and mannose (**a**); galactose, glucose, sucrose, fructose, raffinose, stachyose, verbascose (**b**); corn starch branch chain length distribution (**c**)

4.2.2 GC with Flame Ionization Detection

The gas chromatography (GC) with flame ionization detection has also been extensively used for monosaccharide composition analysis (Chen, Xie, Nie, Li, & Wang, 2008; Yin et al., 2012). However, for this method, monosaccharides need to be derivatized to be volatile before analysis. The commonly used derivatization methods for neutral and acidic sugars are shown in Fig. 4.2. Excellent resolution and robustness are typically associated with GC applications owing to high theoretical chromatographic plates and the inherent purity of the final derived sample (Wang, Liu, Zhou, & Hu, 2012). However, a series chemical reaction including reduction and acetylation are required to convert carbohydrate into volatile compounds, which is complex and time-consuming.

The detailed running condition of alditol acetates in GC-FID can be referred as follows: Using a DB-1701 capillary column (30 m × 0.25 mm × 0.25 μm) and a flame ionization detector, GC is running with an initial column temperature held at 170 °C for 2 min, then programmed at a rate of 10 °C/min to 250 °C, finally held at 250 °C for 10 min (Yin et al., 2012).

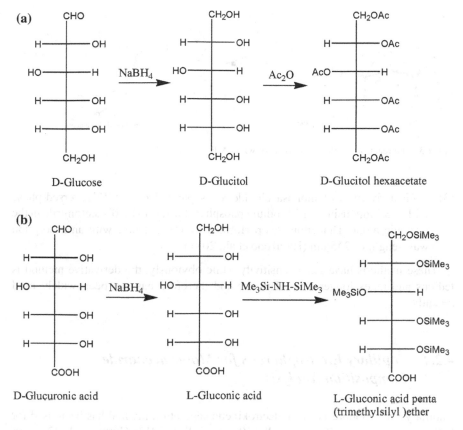

Fig. 4.2 Alditol acetates and TMS derivatives from neutral (**a**) and acidic sugars (**b**), respectively. Adapted from Brummer & Cui, (2005)

4.2.3 Reverse Phase HPLC Methods

High-performance liquid chromatography (HPLC) with fluorometric or UV detection after derivatization with 2-aminopyridine (2-AP), ethyl 4-aminobenzoate (ABEE), 2-aminobenzoic acid (2-AA) or 1-phenyl-3-methyl-5-pyrazolone (PMP) was also extensively used for monosaccharide composition analysis (Harazono et al., 2011). Reverse phase C-18 is commonly used for the separation. Using PMP derivatization as an example (Fig. 4.3), the detailed procedures are as follows:

(1) The dried hydrolyzed polysaccharides are mixed with 50 mL PMP solution in methanol (87.1 mg/mL) and 50 mL 0.3 M sodium hydroxide solution and kept at 70 °C for 30 min.
(2) After neutralization with 0.3 M HCl, the solution is washed once with chloroform to remove the excess reagent.

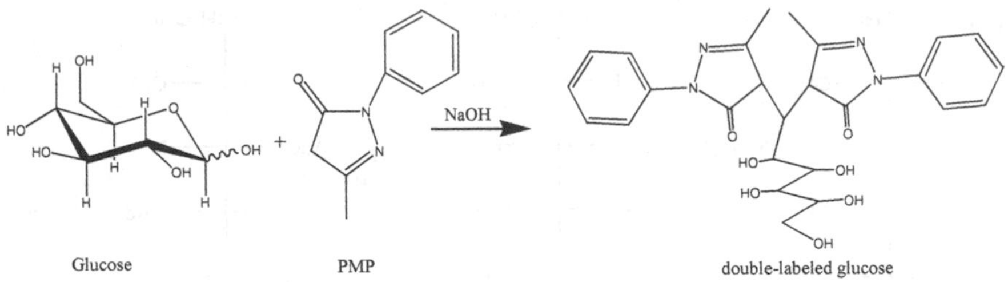

Glucose   PMP   double-labeled glucose

Fig. 4.3 The reaction of glucose derivative with PMP

(3) PMP derivatives of monosaccharide are separated on a C18 reversed-phase
    HPLC column using 0.1 M sodium phosphate buffer (pH 7.0)-acetonitrile in the
    isocratic mode. Detection was performed by UV detector with an absorption
    wavelength of 245 nm (Harazono et al., 2011).

These methods have good sensitivity while obviously, the derivative method is
tedious and takes longer time. This method therefore has not been widely used
recently.

4.2.4 Capillary Electrophoresis for Monosaccharide Composition Analysis

Capillary Electrophoresis as an electrokinetic separation method has been used for
both monosaccharide (Guo, Liu, Jia, Zhang, & Wu, 2013; Guttman, 1997; Starr,
Masada, Hague, Skop, & Klock, 1996) and complex polysaccharides analysis (Volpi,
Maccari, & Linhardt, 2008). The separation principles are as follows: the capillary tube is placed between two buffer reservoirs (Fig. 4.4), and an electric field is
applied. Separation of carbohydrates depends on the electrophoretic mobility and
electro-osmosis of the sugars. For monosaccharide analysis, a bare fused-silica capillary column, laser-induced fluorescence (LIF) detection system and 25 mM lithium
tetraborate buffer (running buffer, pH 10.0) are normally used (Guttman, 1997).

Like reverse phase HPLC method, the pre-separation derivatization of monosaccharides is necessary in order to aid their detection in electrophoresis separations. Many UV active and fluorophore derivatization reagents have been previously suggested including 2-aminopyridine; 4-aminobenzoic acid; 4-aminobenzoate;
4-aminobenzonitrile; 8-aminonaphthalene-1,3,6-trisulfonate (ANTS); 1-phenyl-
3-methyl-2-pyrazolin-5-one (PMP); 8-aminopyrene-1,3,6-trisulfonate(APTS), 9-
aminoacridone (AMAC), etc. The selection of the derivatization reagents are target
dependent. For example, for monosaccharides from neutral sugar and amine sugars,
APTS can be selected for the tagging purpose (Guttman, 1997) (Fig. 4.5).

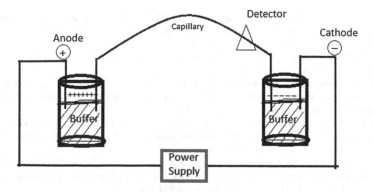

Fig. 4.4 Principles of capillary electrophoresis

Fig. 4.5 Elution profiles of monosaccharides composition analysis using capillary electrophoresis. Peaks **1** AMAC; **2** Neu5Ac-AMAC; **3** GalNAc-APTS; **4** GlcNAc-APTS; **5** Man-APTS; **6** Glc-APTS; **7** Fuc-APTS; **8** Gal-APTS; **9** APTS. Adapted from Guttman, (1997)

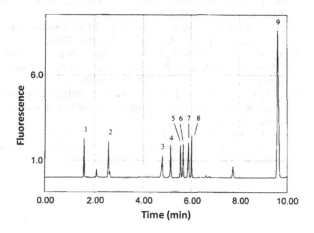

References

Brummer, Y., & Cui, S. W. (2005). Understanding carbohydrate analysis. In S. Cui (Ed.), *Food Carbohydrate: Chemistry, Physical Properties, and Applications* (pp. 67–103). Boca Raton: CRC Press.

Chen, Y., Xie, M.-Y., Nie, S.-P., Li, C., & Wang, Y.-X. (2008). Purification, composition analysis and antioxidant activity of a polysaccharide from the fruiting bodies of *Ganoderma atrum*. *Food Chemistry, 107*(1), 231–241.

Ding, H. H., Cui, S. W., Goff, H. D., Chen, J., Guo, Q., & Wang, Q. (2016). Xyloglucans from flaxseed kernel cell wall: Structural and conformational characterisation. *Carbohydrate Polymers, 151*, 538–545.

Guo, Q., Cui, S. W., Wang, Q., Hu, X., Kang, J., & Yada, R. Y. (2012). Structural characterization of a low-molecular-weight heteropolysaccharide (glucomannan) isolated from Artemisia sphaerocephala Krasch. *Carbohydrate Research, 350*, 31–39.

Guo, Q., Cui, S. W., Wang, Q., Hu, X., Wu, Y., Kang, J., & Yada, R. (2011). Structure characterization of high molecular weight heteropolysaccharide isolated from *Artemisia sphaerocephala* Krasch. seed. *Carbohydrate Polymers, 86*(2), 742–746.

Guo, H. Z., Liu, F. L., Jia, G. Y., Zhang, W. Q., & Wu, F. (2013). Extraction optimization and analysis of monosaccharide composition of fucoidan from *Saccharina japonica* by capillary zone electrophoresis. *Journal of Applied Phycology, 25*(6), 1903–1908.

Guttman, A. (1997). Analysis of monosaccharide composition by capillary electrophoresis. *Journal of Chromatography A, 763*(1–2), 271–277.

Harazono, A., Kobayashi, T., Kawasaki, N., Itoh, S., Tada, M., Hashii, N., …, Yamaguchi, T. (2011). A comparative study of monosaccharide composition analysis as a carbohydrate test for biopharmaceuticals. *Biologicals, 39*(3), 171–180.

Kang, J., Cui, S. W., Chen, J., Phillips, G. O., Wu, Y., & Wang, Q. (2011). New studies on gum ghatti (*Anogeissus latifolia*) part I. Fractionation, chemical and physical characterization of the gum. *Food Hydrocolloids, 25*(8), 1984–1990.

Liu, Y., Zhang, J., Tang, Q., Yang, Y., Guo, Q., Wang, Q., …, Cui, S. W. (2014). Physicochemical characterization of a high molecular weight bioactive β-d-glucan from the fruiting bodies of Ganoderma lucidum. *Carbohydrate Polymers, 101*(0), 968–974.

Starr, C. M., Masada, R. I., Hague, C., Skop, E., & Klock, J. C. (1996). Fluorophore-assisted carbohydrate electrophoresis in the separation, analysis, and sequencing of carbohydrates. *Journal of Chromatography A, 720*(1–2), 295–321.

Volpi, N., Maccari, F., & Linhardt, R. J. (2008). Capillary electrophoresis of complex natural polysaccharides. *Electrophoresis, 29*(15), 3095–3106.

Wang, W., Liu, G., Zhou, B., & Hu, X. (2012). Monosaccharide compositional analysis of purified polysaccharide from *Tricholoma matsutake* by capillary gas chromatography. *Journal of Medicinal Plants Research, 6*(10), 1935–1940.

Wang, Q. C., Zhao, X., Pu, J. H., & Luan, X. H. (2016). Influences of acidic reaction and hydrolytic conditions on monosaccharide composition analysis of acidic, neutral and basic polysaccharides. *Carbohydrate Polymers, 143*, 296–300.

Yin, J. Y., Lin, H. X., Li, J., Wang, Y. X., Cui, S. W., Nie, S. P., & Xie, M. Y. (2012). Structural characterization of a highly branched polysaccharide from the seeds of *Plantago asiatica* L. *Carbohydrate Polymers, 87*(4), 2416–2424.

Chapter 5
Partial Acid Hydrolysis and Molecular Degradation

Polysaccharides are generally degraded into oligosaccharides prior to their structural characterization, while the degradation methods can be classified into two big categories, non-specific degradation and controlled specific degradation.

5.1 Non-specific Degradation

5.1.1 Partial Acid Hydrolysis

Some glycosidic linkages or specific groups in oligosaccharides and polysaccharides are more labile to acid than others; hence, partial acid hydrolysis could help uncover some useful molecular information by understanding the structural features of the hydrolytes afterward. For example, deoxy sugars and furanosyl rings (arabinofuranose) were reported to be easily hydrolyzed by sulphate acid (Guo, Cui, Wang, Hu, Kang & Yada, 2012), while residues with 2-amino-2-deoxyhexose are difficult to be hydrolyzed (Liu, 2005). Polysaccharides with different molecular structure may adopt various partial acid hydrolysis procedures. Mild acids including sulfuric acid, TFA and HCl are frequently used. The temperatures and degradation time should be optimized in the procedure to obtain carbohydrate fragments with varied molecular weights and structural features. For instance, a (galacto)glucomannan structure from ASK seed has been previously reported (Guo et al., 2012), in which the polysaccharides was partially degraded using 0.1 M TFA at 100 °C for 0.5, 1.5, 2.5 and 3.5 h, respectively. After hydrolysis, 3 volumes of ethanol were added to induce precipitation. The precipitates were separated from the supernatant by centrifugation and termed as 0.5P, 1.5P, 2.5P and 3.5P respectively. The HPSEC elution profiles of the hydrolytes and the corresponding monosaccharide composition results are listed in Fig. 5.1 and Table 5.1, respectively.

© Crown 2019
Q. Guo et al., *Methodology for Structural Analysis of Polysaccharides*,
Biobased Polymers, https://doi.org/10.1007/978-3-319-96370-9_5

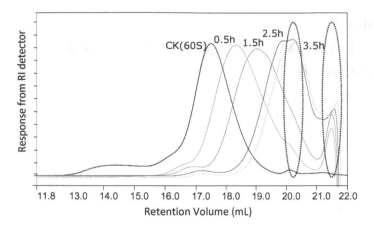

Fig. 5.1 Elution profiles of (galacto)glucomannan and various hydrolysates (a mixture of both large and small molecular weight fragments, denoted with hydrolysis time), adapted from Guo et al. (2012)

Table 5.1 Relative monosaccharide composition for (galacto)glucomannan and its hydrolysates, adapted from Guo et al. (2012)

	Control (%)	0.5P (%)	1.5P (%)	2.5P (%)	3.5P (%)
Ara	9.4	4.8	ND	ND	ND
Gal	24.2	22.2	18.4	13.9	14.5
Glc	39.3	45.9	49.5	49.0	49.4
Man	28.1	33.6	35.8	36.6	36.4

ND: not detected

As can be seen from Table 5.1, the percentage of glucose and mannose increased with the increase of hydrolysis time, which indicated that both sugar residues of glucose and mannose were located in the main chain. This conclusion has been confirmed by the methylation analysis and NMR spectroscopy in this research (Guo et al., 2012).

5.1.2   Ultrasonication Degradation

Ultrasound is a form of energy generating alternative high-pressure (compression) and low-pressure (rarefaction) cycles. It creates cavitation in polysaccharides solution, leading to the degradation of polysaccharide molecules.

The degradation of the polysaccharides is mainly decided by the structural properties. Sonication cleaves polymer in the structurally weakest point. For example, polysaccharides with high Mw and the linear backbone are more easily cleaved comparing to those with lower molecular weight and branched features. The initial molecular weight of the polymers, therefore, plays an important role in the overall

Fig. 5.2 Plot of intrinsic viscosity ([η]) against ultrasonic treatment time for seaweed polysaccharides (from *Porphyra yezoensis* Udea). Concentration and volume of polysaccharide solution: 1.0% (w/v) and 50 mL; parameters of ultrasonic treatment: 20 kHz, 800 W, 30 ± 0.5°C and 4 h. The results are expressed as the mean ± SD (*n* = 3). Adapted from Zhou et al. (2012)

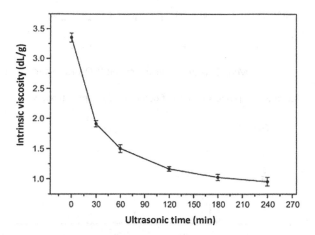

degradation rate. The degradation of the molecular chain has its limit, to which line the chains become so short and stiff that further degradation with ultrasonication time is not possible. However, we can keep increasing the ultrasonication conditions in terms frequency, intensity and temperature to reach another limit of the degradation.

For example, one seaweed polysaccharides *Porphyra yezoensis* Udea (backbone of alternating (1→4)-3,6-anhydro-α-L-galactopyranose units and (1→3)-linked β-D-galactose or (1→4)-linked α-L-galactose 6-sulphate units) were studied for the ultrasonication degradation effects and the results are shown in Fig. 5.2. With the sonication conditions of frequency: 20 kHz, power: 800 W, and temperature: 30 ± 0.5 °C, the intrinsic viscosity of the polysaccharides (1%, w/v) demonstrated a drastically decrease at the first 30 min then gradually reached a plateau when the degradation time reached 4 h (Zhou, Yu, Zhang, He & Ma, 2012).

Ultrasonication has been applied to various polysaccharides, e.g. pectin, carrageenan, guar gum, chitin, chitosan and starch, etc., to decrease the viscosity and/or increase the solubility without affecting the overall structural properties of the molecules (Kulicke, Otto & Baar, 1993; Ogutu, Mu, Elahi, Zhang, & Sun, 2015).

5.2 Controlled Degradation

5.2.1 Enzymatic Degradation

Different from partial acid hydrolysis and ultrasonication degradation, endo-enzymes are able to identify specific linkage types of the polysaccharides and conduct controlled degradation. Although enzymes have been well developed for the hydrolysis of starch, for non-starch polysaccharides degradation, the use of enzyme is largely limited by both the availability of the enzyme and the complex molecular

$$-_4M^1-_4M^1_3-_4M^1-_4M^1-_4M^1_{32}-_4M^1-_4G^1-_4M^1-_4M^1_3-_4M^1_2-_4M^1-_4M^1-_4G^1-_4M^1_2-_4M^1-_4M^1-_4M^1-$$
$$\quad\quad| \quad\quad\quad\quad\quad\quad\quad || \quad\quad\quad\quad\quad\quad | \quad\quad | \quad\quad\quad\quad\quad |$$
$$\quad\quad a \quad\quad\quad\quad\quad\quad\quad aa \quad\quad\quad\quad\quad\quad a \quad\quad a \quad\quad\quad\quad\quad a$$

M: β-D-mannopyranose; G: β-D-glucopyranose; a: O-acetyl group

Fig. 5.3 Proposed structure of dendronan. Adapted from Xing et al. (2015)

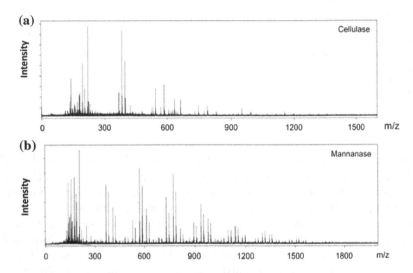

Fig. 5.4 MALDI-TOF-MS profile of acetyl-glucomannan after enzymatic hydrolysis: cellulase (a) and mannanase (b). Adapted from Xing et al. (2015)

structure, although the enzyme degradation has been applied to the degradation of many different polysaccharides, including chitin, galactomannans, mixed-linkage β-glucans, pullulan, xanthan, xylan, acetylxylan, arabinoxylan, cellulose and cellulose derivatives, hemicelluloses, and pectin (BeMiller, 2006). The use of lichenase and β-mannanases for the controlled degradation of β-glucan and galactomannan, respectively has been well summarized by Cui (2005). β-mannanases were applied to the degradation of fenugreek gum, tara gum, guar gum and locust bean gum. Each leads to different oligosaccharides structure, which revealed the different galactose to mannose ratio. The linchenase (E.C. 3.2.1.73) in cereal $(1{\rightarrow}3)(1{\rightarrow}4)$-β-D-glucan target specifically to the $(1{\rightarrow}4)$-linkage of the 3-O-substituted glucose unit of the molecules, which help with the structural characterization.

Using both cellulose and mannase, the degradation of an acetyl-glucomannan structure (Fig. 5.3) from dendrobium officinale has been previously reported (Xing, Cui, Nie, Phillips, Goff, & Wang, 2015).

The generated hydrolyzed fragments were analyzed using MALDI-TOF-MS, which are presented in Fig. 5.4. The detailed mechanisms of MALDI-TOF-MS are explained in the Chap. 8 of this book.

Table 5.2 Methylation analysis results of GSP and its pectinase hydrolytes EGSPP: sugar residues and molar ratios. Adapted from Guo et al. (2015)

RT[a]	Linkage patterns	EGSPP[c]	GSP[c]
0.60	T-Rhap	3.7%	—[d]
0.64	T-Araf	6.5%	3.2%
0.76	T-Fucp	2.4%	–
0.79	T-Arap	2.3%	–
0.95	2-Rhap	5.1%	0.4%
0.98	3-Rhap	2.2%	–
1.00	T-Glcp	2.9%	1.5%
1.05	3-Araf	0.9%	0.2%
1.12	T-GalpA[b]/T-Galp	13.9%	4.5%
1.16	5-Araf	0.8%	0.9%
1.32	2,3-Rhap	1.0%	–
1.37	2,4-Rhap	2.4%	–0.8%
1.39	2,3,4-Rhap	4.4%	–
1.46	3-Galp	7.5%	–
1.53	2-Galp	–	1.6%
1.55	4-GalpA[b]/4-Galp	11.5%	70.7%
1.59	4-Glcp	4.0%	3.8%
1.73	6-Galp	9.0%	1.30%
1.78	3,4- GalpA/Galp	2.5%	2.5%
1.91	2,4-GalpA/Galp	3.8%	1.2%
2.11	4.6-GalpA/Galp	–	1.9%
2.18	3.6- GalpA/Galp	11.2%	4.2%

[a]RT: retention time is relative to 4-Glcp
[b]: dominate percentage in the mixture of GalpA and Galp
[c]Mol (%): molar ratio of each sugar residue is based on the percentage of its peak area
[d] trace amount
EGSPP: the ethanol precipitate of EGSP (pectinase hydrolytes of GSP)

Pectinase was used to help with the structural characterization of pectin (GSP) from ginseng (Guo, Cui, Kang, Ding, Wang, & Wang, 2015). For pectin, some structural units have been defined, including homogalacturonans, rhamnogalacturonan I, and substituted rhamnogalacturonan II (Yu, Lutterodt, & Cheng, 2009). The existence of sugar residues 2-Rhap and 2,4-Rhap together with the presence of galactan and arabinan sugar residues confirmed the presence of rhamnogalacturonan I structural character in GSP molecules. However, due to the small percentage of above sugar residues in GSP (Table 5.2), the sequences of arabinose and galactose-based

Fig. 5.5 Elution profiles of ginseng polysaccharides (GSP) and its fractions (EGSP: GSP after pectinase hydrolysis; EGSPP: the precipitate of EGSP after ethanol precipitation process; EGSPS: the supernatant of EGSP after ethanol precipitation process) using HPSEC coupled with RI detector. Adapted from Guo et al. (2015)

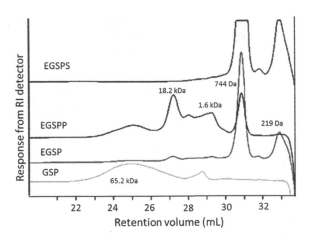

sugar residues in GSP cannot be determined even using NMR spectroscopy, enzyme hydrolysis combined with ethanol precipitation, therefore, were used to solve this problem.

In our research (Guo et al., 2015), GSP (10 mg/mL) was completely hydrolyzed by pectinase (Sigma, P1746) at pH 4.5, temperature 50 °C for 24 h under constant stirring (dilute HCl and NaOH solution was used during this process to control the pH). The GSP hydrolysates (EGSP) solution was then mixed with three volumes of ethanol to obtain EGSP precipitate (EGSPP) and EGSP supernatant (EGSPS), which are presented in Fig. 5.5 (Guo et al. 2015).

As can be seen from the sugar residues information in Table 5.2, EGSPP contained higher amount of rhamnose and galactose compared to GSP, which are derived from the hairy region of GSP (rhamnogalacturonan structure and other branching regions), as the branching structure could block the access of enzyme, therefore can be recovered through ethanol precipitation.

5.3 Removal of Glycosides from Glycoproteins for Structure Determination

For glycoprotein, the glycosides need to be detached from protein before structural characterization. This has been previously summarized by Robyt (1998). For example, the O-glycosides attached to serine and threonine and S-glycosides attached to cysteine can be simply released by mild alkaline treatment through β-elimination, while the sodium borohydride is usually used to prevent the degradation of glycosides. The starting reaction conditions can be referred to follows: 0.1 M NaOH and 0.3 M NaBH$_4$ at 37 °C for 48 h. However, the conditions should be adjusted for different types of glycoprotein or glycopeptides.

For C- or N-terminal residue, the β-elimination cannot satisfy the requirement anymore. Other methodologies such as hot anhydrous hydrazine can be used for asparagine N-linked glycosides hydrolysis (hydrazinolysis), endoglycosidases such as endo-β-*N*-acetylglucosaminidase can also be used to liberate asparagine N-linked oligosaccharides chain (Takasaki, Mizuochi, & Kobata, 1982).

Trifluoromethanesulfonic acid (TFMS) as chemical methods could remove both O- and N-linked oligosaccharides from glycoproteins and leave an intact protein component and released oligosaccharides fragments for further analysis. For TFMS hydrolysis, sample need to be placed in an very cold temperature. For example, ethanol/dry ice bath is normally required. The detailed procedures (Sojar & Bahl, 1987) of TFMS hydrolysis are addressed in Chap. 10 of this book.

References

BeMiller, J. (2006). Gums and hydrocolloids. In *Carbohydrates in food* (2nd ed., pp. 209–231). CRC Press

Cui, S. (2005). Structural analysis of polysaccharides. In S. Cui (Ed.). *Food carbohydrates: Chemistry, physical properties, and applications* (pp. 106–160). CRC Press

Guo, Q. B., Cui, S. W., Kang, J., Ding, H. H., Wang, Q., & Wang, C. (2015). Non-starch polysaccharides from American ginseng: Physicochemical investigation and structural characterization. *Food Hydrocolloids, 44*, 320–327.

Guo, Q., Cui, S. W., Wang, Q., Hu, X., Kang, J., & Yada, R. Y. (2012). Structural characterization of a low-molecular-weight heteropolysaccharide (glucomannan) isolated from Artemisia sphaerocephala Krasch. *Carbohyd Res, 350*, 31–39.

Kulicke, W.-M., Otto, M., & Baar, A. (1993). Improved NMR characterization of high-molecular-weight polymers and polyelectrolytes through the use of preliminary ultrasonic degradation. *Die Makromolekulare Chemie, 194*(3), 751–765.

Liu, Q. (2005). Understanding starches and their role in foods. In S. W. Cui (Ed.), *Food carbohydrates: Chemistry, physical properties, and applications. Taylor and Francis, Boca Raton, Florida* (pp. 309–355). CRC Press

Ogutu, F. O., Mu, T. H., Elahi, R., Zhang, M., & Sun, H. N. (2015). Ultrasonic modification of selected polysaccharides-review. *Journal of Food Processing & Technology, 6*(446)

Robyt, J. F. (1998). Determinations. In *Essentials of carbohydrate chemistry* (pp. 345–364). New York, NY: Springer New York

Sojar, H. T., & Bahl, O. P. (1987). A chemical method for the deglycosylation of proteins. *Arch Biochem Biophys, 259*(1), 52–57.

Takasaki, S., Mizuochi, T., & Kobata, A. (1982). Hydrazinolysis of asparagine-linked sugar chains to produce free oligosaccharides. *Methods Enzymol, 83*, 263–268.

Xing, X., Cui, S. W., Nie, S., Phillips, G. O., Goff, H. D., & Wang, Q. (2015). Study on *Dendrobium officinale O*-acetyl-glucomannan (Dendronan®): Part V. Fractionation and structural heterogeneity of different fractions. *Bioactive Carbohydrates and Dietary Fibre, 5*(2), 106–115

Yu, L., Lutterodt, H., & Cheng, Z. (2009). Beneficial health properties of psyllium and approaches to improve its functionalities. In S. L. Taylor (Ed.), *Advances in food and nutrition research* (Vol. 55, pp. 193–220)

Zhou, C., Yu, X., Zhang, Y., He, R., & Ma, H. (2012). Ultrasonic degradation, purification and analysis of structure and antioxidant activity of polysaccharide from Porphyra yezoensis Udea. *Carbohydrate Polymers, 87*(3), 2046–2051.

Chapter 6
Linkage Pattern Analysis

6.1 Methylation Analysis

Methylation analysis, a powerful tool for primary structure characterization, has been used for decades and is still widely used nowadays (Chandra, Ghosh, Ojha, & Islam, 2009; Pereira et al., 2010; Wu, Cui, Eskin, & Goff, 2009). As shown in Fig. 6.1, it includes four main steps: (1) methylation reaction using methyl iodide to convert all free hydroxyl groups of the polysaccharide molecules into methoxy groups; (2) acidic hydrolysis (usually by TFA) to convert the polymer into monomer; (3) reduction (using sodium borodeuteride) and acetylation (by acetic anhydride) to give volatile products: partially methylated alditol acetates (PMAA); (4) GC-MS analysis to identify and quantify the produced PMAAs (Dell, 1990; Liu, 2005). The linkage patterns for each monomer and the molar ratios can be obtained using this method (Ciucanu & Kerek, 1984).

The electron-impact fragmentation patterns of the mass spectra of PMAAs (Fig. 6.1) are well documented for all linkage patterns and for all known sugars by Carpita and Shea (1989). The substitution pattern of PMAAs can be readily determined based on the following rules (Cui, 2005):

Rule 1: Only the alditol backbone was fragmented;

Rule 2: The charge always resides on the fragment with a methoxy-bearing carbon atom adjacent to the cleavage point;

Rule 3: Fragmentation between two adjacent methoxy-bearing carbon atoms is favored over fragmentation between a methoxy-bearing carbon atom and an acetoxy carbon atom, which itself is highly favored over fragmentation between two acetoxy-bearing carbon atoms;

Rule 4: Secondary fragment-ions are produced by the loss of methanol or acetic acid. The carbon from which the substituted group is cleaved is preferred to bear the charge;

© Crown 2018
Q. Guo et al., *Methodology for Structural Analysis of Polysaccharides,*
Biobased Polymers, https://doi.org/10.1007/978-3-319-96370-9_6

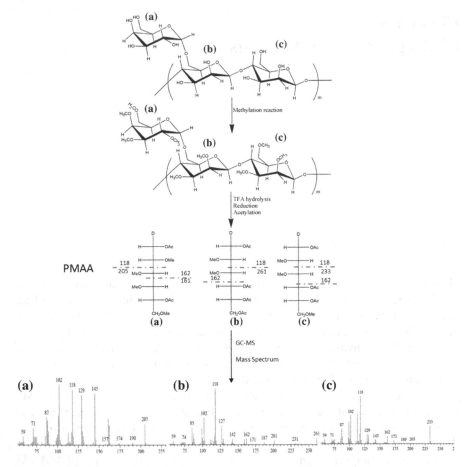

Fig. 6.1 Methylation analysis of galactomannan structure and the corresponding mass spectrum, **a** t-gal*p*, **b** 4,6-man*p*, **c** 4-man*p*

Rule 5:   When the PPMA is labeled at C-1 with a deuterium atom, the charge-to-mass ratio (m/z) of a fragment ion that contains C-1 is even, whereas m/z of a fragment ion that does not contain C-1 is odd.

However, the mass spectrum database for all possible PMAAs is also currently reachable online (https://www.ccrc.uga.edu/specdb/ms/pmaa/pframe.html#na).

The estimated molar ratio of each sugar residue can be determined using the percentage of the peak area. The molar ratio obtained, however, is only an estimated value, as different sugar residues have distinct peak response factors. Therefore, it should always be confirmed with the results from both monosaccharide composition analysis and ^{1}H NMR spectrum. It should be noted that considerable variation exists in the rates of hydrolysis after methylation, as the reactive anomeric carbons are more susceptible to degradation during hydrolysis, especially the furanosyl linkages. Therefore, some loss of arabinosyl and/or fructofuranosyl residues is expected (Carpita & Shea, 1989).

The degree of branching (DB) of polysaccharides can be calculated using the data from methylation analysis according to the equation below (Qian, Cui, Nikiforuk, & Goff, 2012):

$$DB = (N_T + N_B)/(N_T + N_B + N_L)$$

where NT, NB, and NL are the molar percentage of the terminal, branched, and linear residues, respectively. The DB value of a linear chain equals 0, whereas that of a fully branched polymer is 1.

6.2 Acidic Polysaccharides Reduction Before Methylation Analysis

The high percentage of uronic acids in acidic polysaccharides, e.g. pectin, created difficulties for methylation analysis which in most cases generates less or even no PMAAs. Therefore, uronic acid firstly need to be reduced into neutral sugars prior to methylation analysis. The related steps (Fig. 6.2) are described as follows (Taylor & Conrad, 1972; York, Darvill, McNeil, Stevenson, & Albersheim, 1986).

(1) Sample preparation and activation

The acidic polysaccharide is dissolved in deuterium oxide (D_2O). To the solution, certain amount of 1-cyclohexyl-3-(2-morpholinoethyl)-carboiimidemethyl-*p*-toluenesulfonate (CMC) is added under the pH of 4.75 and then left for 1 h for sample activation.

(2) Reduction

The polysaccharide is then reduced by dropwise adding adequate amount of sodium borodeuteride solution under pH 7. The reduced polysaccharide is separated from salts by dialysis against distilled water and then lyophilized.

(3) Sample purification and preparation for the methylation analysis

The polysaccharide is re-dissolved in 1 mL distilled water and 10% acetic acid in methanol is added. The mixture is dried under a stream of nitrogen to remove the boric acid. Another 1 mL of 10% acetic acid in methanol is added to the residue and evaporated using nitrogen. This process need to be repeated 3–4 times to ensure that most of the boric acid is removed. Finally, a few drops of methanol are added and the solution was evaporated (two times) to remove any boric acid remaining.

It should be noted that sodium borodeuteride ($NaBD_4$) instead of sodium boro-hydride ($NaBH_4$) should be used to reduce the polysaccharides. The –COOH group can be reduced to CD_2OH instead of CH_2OH, which can be distinguished by the

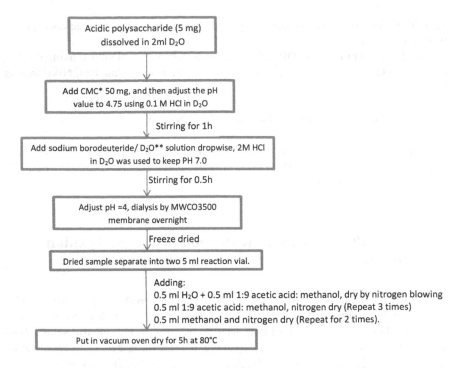

* CMC: 1-Cyclohexyl-3-(2-morpholinoethyl)carbodiimide metho-p-toluenesulfonate

** 800 mg sodium borodeuteride dissolved into 5 mL D₂O

Fig. 6.2 Procedures of pectic polysaccharides reduction before methylation analysis

Fig. 6.3 PMAA derived
from T-Glc*p* (a) and T-Glc*p*A
after NaBD₄ reduction (b)

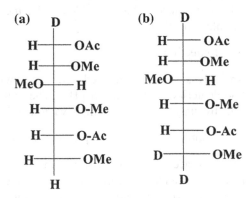

PMAA structure from the mass spectrum (Fig. 6.3). For example, in comparison with T-Glc*p*, MS spectrum of PMAA derived from reduced T-Glc*p*A contains peaks such as 147, 163 and 207 instead of 145, 161 and 205.

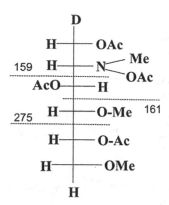

Fig. 6.4   Structure of 2-deoxy-2-N-methylacetamidohexitol acetates

6.3   Strategies of Methylation Analysis for Glycoprotein or Other Uncommon Sugars

For amino sugars, successful permethylation of complex carbohydrates containing a 2-deoxy-2-acetamidosugar is accompanied by N-methylation of an acetamido group, which lead to highly resistance to acid hydrolysis for sugar residues bearing this groups, resulting in the failure of methylation analysis. A modified procedure such as acetolysis or methanolysis needs to be conducted. For example, after permethylation, amino sugar need to be treated with 0.5 M sulfuric acid in 95% acetic acid at 80 °C under stirring overnight. The reaction mixture was then mixed with 0.3 ml water and heated at 80 °C for an additional 5 h to produce 2-deoxy-2-N-methylacetamidohexitol acetates, which can be identified by GC-MS by the specific mass peak of 159, 275 (Stellner, Saito et al., 1973) as shown in Fig. 6.4.

For sugars which are endogenously methylated, the position of these groups can be revealed by the use of CD_3I or CH_3CH_2I instead of CH_3I, which can be quantified by GC-MS; positions of endogenous acetylation can also be monitored by chemical derivatization in reactions compatible with the methylation analysis (Carpita & Shea 1989)

6.4   Smith Degradation (Goldstein, Hay, Lewis, & Smith, 1959, 1965)

The Smith degradation is helpful to understand the basic structure information of complex oligosaccharides or polysaccharides. Three main steps are involved in this approach: periodate oxidation, reduction and mild acid hydrolysis. After periodate oxidation, a dialdehyde and formic acid can be formed. The dialdehyde could be

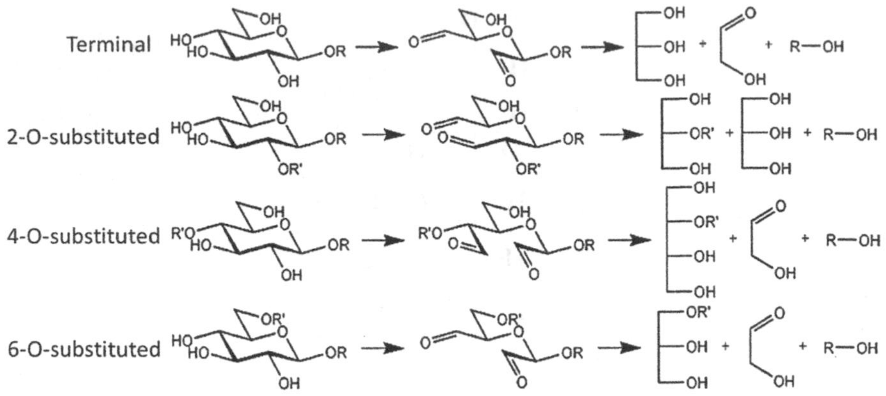

Fig. 6.5 Adapted from (http://www.stenutz.eu/sop/sop101.html)

converted to the corresponding alcohols by borohydrides reduction. The end product (acetals) are ready to be hydrolyzed even with dilute acid at room temperature (Lindberg, Lönngren, & Svensson, 1975).

6.4.1 Periodate Oxidation

Free two adjacent hydroxyl groups (2-linked and 4-linked residue) in the sugar residue can be oxidized by periodic acid to form two aldehydic groups upon the cleavage of the carbon chain, in which one molar equivalent of periodate will be consumed. In the case of three adjacent hydroxyl groups (terminal and 6-linked residue), the oxidation will consume two molar equivalents of periodate, at the same time, one molar equivalent of formic acid will be generated. For 3-linked sugar residue, no periodate will be consumed due to the lack of adjacent free hydroxyl group in sugar residue. The periodate oxidation, therefore, can be used to deduce the linkage pattern of sugar residue. It could also be used to estimate the degree of polymerization of linear $1\rightarrow4$ or $1\rightarrow2$-linked polysaccharides, as well as to estimate the degree of branching by calculating the ratio of the terminal to non-terminal sugar residues of a branched polysaccharide.

6.4.2 Reduction and Acid Hydrolysis

After oxidation, the resultant dialdehyde could be reduced with borohydrides to give the corresponding alcohols (Fig. 6.5), which is a true acetal and sensitive to acid. These acetals are ready to be hydrolyzed even with dilute acid at room temperature. In contrast, the surviving sugar units in glycosidic linkages are resistant to hydrolysis.

References

Carpita, N. C. and Shea, E. M. (1989) Linkage structure of carbohydrates by gas chromatography-mass spectrometry (GC-MS) of partially methylated alditol acetates, in Analysis of Carbohydrates by GLC and MS (C. J. Biermann and G. D. McGinnis, eds.), CRC Press, Boca Raton, FL, p. 157.

Chandra, K., Ghosh, K., Ojha, A. K., & Islam, S. S. (2009). Chemical analysis of a polysaccharide of unripe (green) tomato (Lycopersicon esculentum). *Carbohydrate Research, 344*(16), 2188–2194.

Ciucanu, I., & Kerek, F. (1984). A simple and rapid method for the permethylation of carbohydrates. *Carbohydrate Research, 131*(2), 209–217.

Cui, S. (2005). Structural analysis of polysaccharides. In S. Cui (Ed.), *Food carbohydrates: Chemistry, physical properties, and applications* (pp. 106–160). CRC Press.

Dell, A. (1990). Preparation and desorption mass spectrometry of permethyl and peracetyl derivatives of oligosaccharides. *Methods in Enzymology, 193*, 647–660.

Goldstein, I. J., Hay, G. W., Lewis, B. A., & Smith, F. (1959). *A new approach to the determination of the fine structure of polysaccharides. Abstracts, 135th Meeting American Chemical Society D* (Vol. 3).

Goldstein, I. J., Hay, G. W., Lewis, B. A., & Smith, F. (1965). Controlled degradation of polysaccharides by periodate oxidation, reduction, and hydrolysis. *Methods Carbohydr. Chem, 5*, 361–370.

Guo, Q., Cui, S. W., Wang, Q., Hu, X., Wu, Y., Kang, J., et al. (2011). Structure characterization of high molecular weight heteropolysaccharide isolated from Artemisia sphaerocephala Krasch seed. *Carbohydrate Polymers, 86*(2), 742–746.

Lindberg, B., Lönngren, J., & Svensson, S. (1975). Specific degradation of polysaccharides. *Advances in Carbohydrate Chemistry and Biochemistry, 31*, 185–240.

Liu, Q. (2005). Understanding Starches and Their Role in Foods. In S. W. Cui (Ed.), *Food carbohydrates: Chemistry, physical properties, and applications*. Taylor and Francis, Boca Raton, Florida (pp. 309–355). CRC Press.

Pereira, M. I., Ruthes, A. C., Carbonero, E. R., Marcon, R., Baggio, C. H., Freitas, C. S., et al. (2010). Chemical structure and selected biological properties of a glucomannan from the lichenized fungus Heterodermia obscurata. *Phytochemistry, 71*(17–18), 2132–2139.

Qian, K.-Y., Cui, S. W., Nikiforuk, J., & Goff, H. D. (2012). Structural elucidation of rhamnogalacturonans from flaxseed hulls. *Carbohydrate Research, 362*, 47–55.

Stellner, K., Saito, H., et al. (1973). Determination of aminosugar linkages in glycolipids by methlation -aminosugar linkages of ceramide pentasaccharides of rabbit erythrocytes and of forssman antigen. *Archives of Biochemistry and Biophysics, 155*(2), 464–472.

Sweet, D. P., Albersheim, P., & Shapiro, R. H. (1975). Partially ethylated alditol acetates as derivatives for elucidation of the glycosyl linkage-composition of polysaccharides. *Carbohydrate Research, 40*(2), 199–216.

Taylor, R. L., & Conrad, H. E. (1972). Stoichiometric depolymerization of polyuronides and glycosaminoglycuronans to monosaccharides following reduction of their carbodiimide-activated carboxyl groups. *Biochemistry, 11*(8), 1383–1388.

Wu, Y., Cui, W., Eskin, N. A. M., & Goff, H. D. (2009). Fractionation and partial characterization of non-pectic polysaccharides from yellow mustard mucilage. *Food Hydrocolloids, 23*(6), 1535–1541.

York, W. S., Darvill, A. G., McNeil, M., Stevenson, T. T., & Albersheim, P. (1986). Isolation and characterization of plant cell walls and cell wall components. In H. W. Arthur Weissbach (Ed.), *Methods in enzymology* (Vol. 118, pp. 3–40). Academic Press.

Chapter 7
1D & 2D and Solid-State NMR

NMR spectroscopy, as a powerful tool for carbohydrate structural analysis, could provide a complete picture of oligosaccharides structure and their behavior in solution. Structural features including linkage patterns, configuration (α- or β-), sequencing as well as the ratio of different sugar residues all could be obtained. In addition, some conformational properties (molecular dynamics) of polysaccharides also are obtainable by this technique.

The detailed information for 1D and 2D NMR analysis for oligosaccharides has been thoroughly addressed in several review papers (Agrawal, 1992; Bubb, 2003), while this chapter mainly focused on the framework of how stepwise the chemical shifts of each sugar residues can be assigned as well as how the sequence of residues can be established.

7.1 Sample Preparation

Prior to NMR testing, polysaccharides need to be dissolved in D_2O or DMSO-d_6 to convert the entire free –OH into –OD as shown in the Fig. 7.1. To complete the conversion, three cycles of "dissolving in D_2O + freeze drying" is necessary. Polysaccharides sample should be well dissolved; any solid particles suspended in the solution could distort the field homogeneity and lead to a broad peak. Therefore, filtration sometimes is required before transferring the sample into the NMR tube. In case of the sample with high viscosity, the relatively high temperature is necessary to bring down the viscosity during the testing.

© Crown 2018
Q. Guo et al., *Methodology for Structural Analysis of Polysaccharides*,
Biobased Polymers, https://doi.org/10.1007/978-3-319-96370-9_7

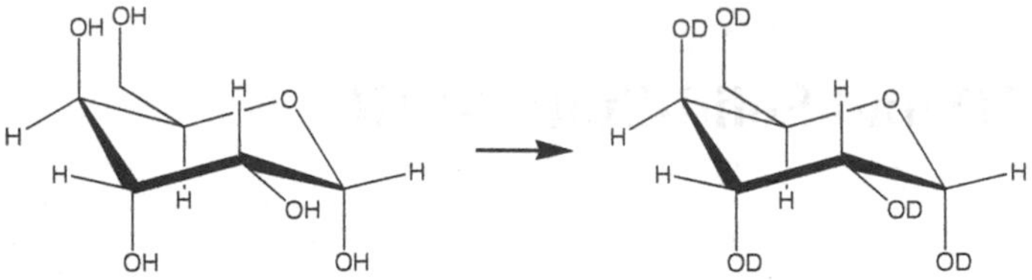

Fig. 7.1 The conversion of –OH into –OD to eliminate the proton peaks derived from –OH

7.2   1D NMR

The most commonly used 1D NMR for polysaccharides characterization including
^{1}H and ^{13}C NMR. For proton NMR, three types of information can be extracted for
structural analysis: chemical shift, peak intensity (integration) and spin-spin coupling
distance. For polysaccharides, the signals are mostly overlapped and are difficult to
be assigned, as most of the proton signals fall within a 2 ppm chemical shift range
(3–5 ppm), which results in substantial overlap of the signals. However, some well-
resolved signals, including those of anomeric protons (4.4–5.5), acetyl (2.0–2.1)
and methyl (1.2 ppm) groups, and other protons that are influenced by functional
groups existed, which add value to the ^{1}H NMR (Fig. 7.4a). In addition, the α- and β-
configurations for most of the sugar residues could be easily identified. For example,
most of the α-anomeric protons will appear in the region of 5–6 ppm while most
of the β-anomeric protons will appear in the 4–5 ppm range. Compared to ^{13}C, ^{1}H
NMR signals are much more sensitive due to their natural abundance. As a result, ^{1}H
NMR signals can be used for quantitative purposes: the molar ratio of sugar residues
as well as the substitution of some functional groups could also be manifested by
the peak integration from ^{1}H NMR. As shown in Fig. 7.2, the H1 of four sugar
residues including A: T-α-D-glcpA; B: 4-α-D-galpA; C: 2,4-β-D-xylp; D: 4-β-D-xylp
are demonstrated in the anomeric region, from which C, D with the chemical shift
of 4.61 and 4.44 ppm contained β-configuration and A, B with chemical shift above
5 ppm have α-configuration. The integration of each peak can be used to determine
the abundance of the corresponding sugar residues in this polysaccharides, which
can be compared with data from both monosaccharide composition analysis and
methylation analysis.

^{13}C NMR provides well-resolved peaks compared to ^{1}H NMR due to its wider
chemical shift dispersion and lack of complexities arising from spin-spin coupling
and overlap of resonances with those arising from solvents (Agrawal, 1992), although
the abundance is lower. For example (Fig. 7.4b), anomeric carbons mostly appear
in the 90–110 ppm while the non-anomeric carbons are between 60 and 85 ppm.
As for polysaccharides with de-oxygen sugars such as rhamnose, the –CH$_3$ signals
appear in 15–20 ppm. Sugar residues with α-anomeric carbons appear in the region
of 95–101 ppm while most of the β-anomeric carbons appear between 101 and
105 ppm. Exceptions appeared when dealing with mannose and rhamnose, where
the C1 chemical shift is uninformative in establishing anomeric configuration. The

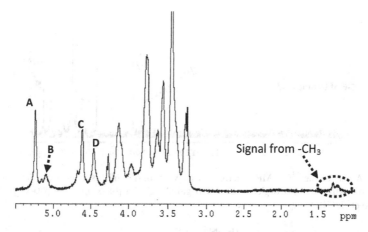

Fig. 7.2 A part of the ¹H NMR spectrum of polysaccharide 60P isolated from ASK (60 °C in D₂O). From which A: T-α-D-glc*p*A; B: 4-α-D-gal*p*A; C: 2,4-β-D-xyl*p*; D: 4-β-D-xyl*p*. Adapted from Guo et al., (2011)

chemical shifts of C3 and C5 are of diagnostic importance because these absorb at 1.5–3.0 ppm higher field in the α-anomer relative to the β-anomer. The carbon atoms that are bearing the primary hydroxyl groups including C6 in pyranoses and C5 in furanoses, showed signals at the relative high field of 60–64 ppm, while the signals of carbon atoms with secondary hydroxyl groups (C2,3,4 in pyranoses and C2,3 in furanoses) appear in the region of 65–85 ppm. Carboxyl carbons derived from uronic acid and/or the carbonyl group of acetamido sugars show the signal in a much lower field, i.e., 170–180 ppm. The glycosylation shift also influences its adjacent carbons which tend to shift up-fields by a small amount (0.9–4.6 ppm) (Fig. 7.4b).

As shown in Fig. 7.3, well-resolved peaks were detected in the ¹³C spectrum. C1 of residue **C** and **D** overlapped at 102 ppm while residue **A** and **B** overlapped at 98 ppm, the C-6 of uronic acid of A/B showed peaks at 178 ppm. Due to the low abundance of ¹³C in the molecules, the peak integration is usually not used for the quantification of the sugar residues (Fig. 7.4).

7.3 2D NMR

Due to the limitation of 1D NMR and the complexity of the molecular structure of polysaccharides, the complete chemical shift assignment is nearly impossible based on the 1D NMR only. 2D NMR, which demonstrates the correlations between protons (COSY, TOCSY, NOESY) and between proton and carbons (HMQC, HMBC), can help with the ¹H and ¹³C chemical shift assignments and provide critical structural information such as sequence of the various sugar residues.

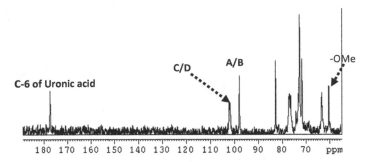

Fig. 7.3 A part of the ^{13}C NMR spectrum of polysaccharide 60P isolated from ASK (60 °C in D$_2$O). From which A: T-α-D-glcpA; B: 4-α-D-galpA; C: 2,4-β-D-xylp; D: 4-β-D-xylp. Adapted from Guo et al. (2011)

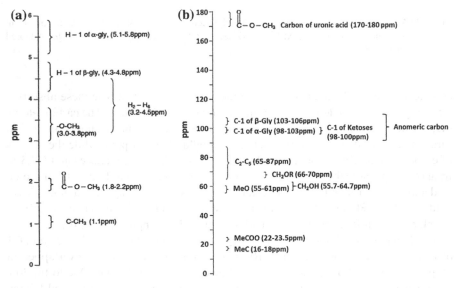

Fig. 7.4 Chemical shifts of carbohydrate in ^{1}H (**a**) and ^{13}C (**b**). Adapted from Cui (2005a)

7.3.1 COSY Spectrum

COSY is short for correlation spectroscopy, which provides the correlation of the adjacent proton in one sugar ring. The correlation is determined by the spin-spin coupling between protons. COSY spectrum contains a diagonal and cross peaks. The general approach is to start with an isolated resonance, often an anomeric proton (4.3–5.4 ppm) or the methyl resonance (1.2–1.4 ppm) in 6-deoxy sugars (rhamnose and fucose), then to correlate spins in a step-wise manner around the spin system of the ring. As shown in Fig. 7.5, for sugar residue A (T-GlcpA), the starting resonance was H1 (5.22 ppm), then the chemical shift of H2 (3.58 ppm), H3 (3.75 ppm), H4

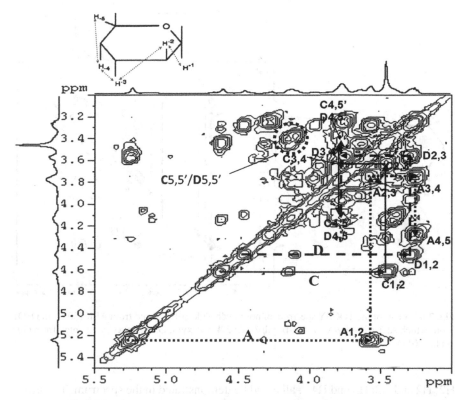

Fig. 7.5   A part of the COSY spectrum of polysaccharide 60P isolated from ASK (60 °C in D$_2$O). Proton correlations of sugar residues A, C and D were labeled. From which A: T-α-D-glc$p$A; B: 4-α-D-gal$p$A; C: 2,4-β-D-xyl$p$; D: 4-β-D-xyl$p$. Adapted from Guo et al., (2011)

(3.26 ppm) and H5 (4.26 ppm) were all obtained through the scalar connectivity. In the similar manner, the complete chemical shifts of residues C (2,4-β-D-xyl$p$) and D (4-β-D-xyl$p$) were obtained and labeled in Fig. 7.5. However, it should be brought to our attention that the small or no couplings do exist in adjacent protons of some sugar residues. For example, the coupling between H4 and H5 (J 4, 5 = 2–3 Hz) of a galactopyranosyl residue and coupling between H1 and H2 in a mannopyranosyl residue, prevents detection of cross peaks in H–H correlation spectrum, which creates difficulties for complete proton assignments.

7.3.2   TOCSY Spectrum

TOCSY (total correlation spectroscopy) also known as homonuclear Hartmann-Hahn spectroscopy (HOHAHA), provides the total correlation of protons within one sugar ring (referred as J-network) (Agrawal, 1992), e.g. the correlation between H1 and

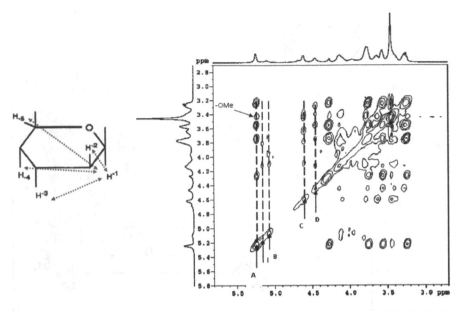

Fig. 7.6 A part of the TOCSY spectrum of polysaccharide 60P isolated from ASK (60 °C in D$_2$O). From which A: T-α-D-glc$p$A; B: 4-α-D-gal$p$A; C: 2,4-β-D-xyl$p$; D: 4-β-D-xyl$p$. Adapted from Guo et al., (2011)

H2; H2 and H3; H3 and H4… all could be demonstrated in the spectrum. Therefore, you may notice peaks for each sugar residue form one line in TOCSY spectrum (Fig. 7.6). TOCSY is especially useful as a complementary method of COSY for the complete proton assignments. However, the chemical shift of sugar residues from references are always of great help for the proton assignments.

7.3.3 HSQC Spectrum

The correlations between proton and the directly linked ^{13}C could be demonstrated by heteronuclear multiple-quantum coherence (HMQC) or heteronuclear single quantum coherence (HSQC). The correlation of anomeric ^{1}H and ^{13}C is well recognized compared to the non-anomeric connections. As shown in Fig. 7.7, the correlation between anomeric carbon and proton of sugar residues A, B, C and D are well separated in the circled region, while interactions (peaks) of H2/C2 to H5/C5 are all noticeable from non-anomeric region (Fig. 7.7).

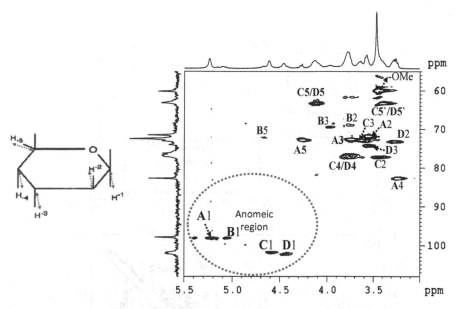

Fig. 7.7   A part of the HSQC spectrum of polysaccharide 60P isolated from seeds of ASK (60 °C in D$_2$O). A: T-α-D-glc$p$A; B: 4-α-D-gal$p$A; C: 2,4-β-D-xyl$p$; D: 4-β-D-xyl$p$. Adapted from Guo et al., (2011)

7.3.4   Sequence Determination by NOESY and HMBC

Successful assignments of all the peaks in COSY, TOCSY and HSQC spectra could deliver the complete ^{1}H and ^{13}C chemical shifts of all possible sugar residues. The sequences of the residues then need to be determined for a proposed molecular structure of polysaccharides. For which the long-range ^{1}H and ^{13}C correlation (HMBC) and ^{1}H and ^{1}H correlation (NOESY) could be used to solve this problem.

The heteronuclear multiple bond connectivity (HMBC) spectroscopy could provide long-range coupling between proton and carbon (two or three bonds distance) with great sensitivity (Fig. 7.8a). It not only confirms ^{13}C and/or ^{1}H assignments, but also provides multiple bond correlations between the anomeric carbon and the aglycone proton or between the anomeric proton and the aglycone carbon, and thus able to identify inter-glycosidic linkage, as shown in Fig. 7.9a (Agrawal, 1992). The sequences of the sugar residues therefore can be provided, which stepwise lead to information about the skeleton of a molecule.

Nuclear overhauser effect spectroscopy (NOESY) provides information *through-space* rather than *through-bond* couplings. NOE connectivities are often observed between the anomeric proton of a particular sugar residue to protons of the other sugar residue that is glycosidically linked to the former (Fig. 7.8b). The inter-residue NOE could provide sequence information of a polysaccharide or oligosaccharide (Fig. 7.9b). NOESY is one of the most useful techniques as it allows one to correlate

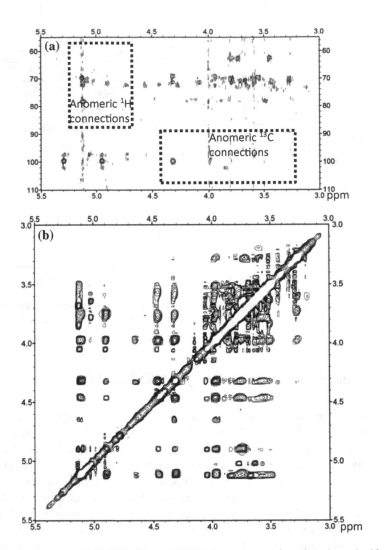

Fig. 7.8 Key fragment of HMBC (**a**) and NOESY (**b**) spectrum of pectic polysaccharides from flaxseed hulls. Adapted from Qian, Cui, Nikiforuk, & Goff, (2012)

nuclei through space (distance smaller than 5 Å) and the distance between two protons can be extracted by measuring cross peak intensity. Thus NOE experiments have been used for measuring the conformation of carbohydrates (Cui, 2005a).

In both HMBC and NOESY spectra, numerous peaks derived from inner residue coupling which are uninteresting for assigning the glycosidic linkages exsit. Therefore, Comparisons of HMBC with HSQC and NOESY with TOCSY spectra are

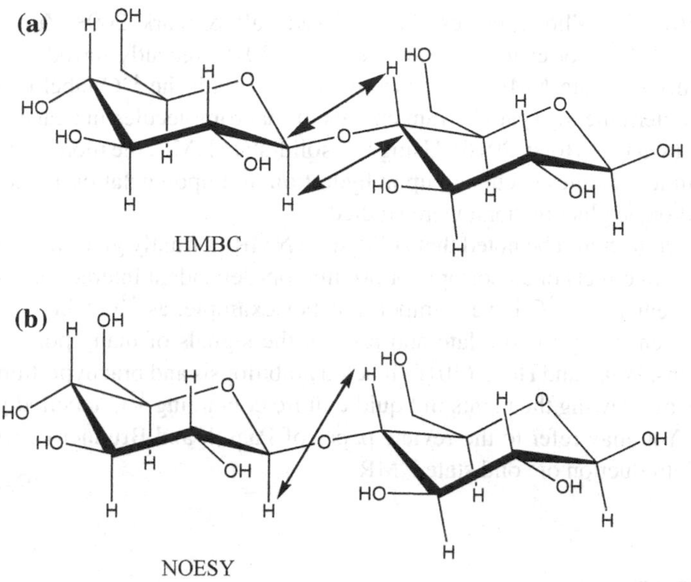

Fig. 7.9 Interglycosidic connectivity observed by HMBC (**a**) and NOESY (**b**)

necessary and helpful to assign peaks and extract information in need. It also should be noted that relatively weak peaks are obtained for inter-residues correlations due to the long-range or multi-bond coupling, which creates difficulties for the determination of the sequences between residues.

7.4   Solid-State NMR

The normal NMR analysis of polysaccharides is solution based. Polysaccharides must dissolve in deuterated water, DMSO, etc. prior to NMR analysis. Some polysaccharides need to be modified first (methylated) to enhance their solubility in the solvent mentioned above, which increases the difficulties of peak assignments in the spectrum. Some polysaccharides, such as cellulose, are still difficult to be dissolved even after modification. In this situation, solid-state NMR demonstrates its advantages. It should be noted that this technique has been applied to the characterization of polysaccharides for quite a long time (Bootten, Harris, Melton, & Newman, 2009; Dick-Perez, Wang, Salazar, Zabotina, & Hong, 2012; Dick-Perez, Zhang, Hayes, Salazar, Zabotina, & Hong, 2011; Ivanir & Goldbourt, 2014; Ng, Zujovic, Smith, Johnston, Schroeder, & Melton, 2014; Pizzoferrato, Manzi, Bertocchi, Fanelli, Rotilio, & Paci, 2000; Redenti et al., 1999; Spevacek & Brus, 2008; Tzou, 2005; Vieira & Gil, 2005; Wawer, Wolniak, & Paradowska, 2006; White, Wang, Park, Cosgrove, & Hong, 2014; Zhang, Klymachyov, Brown, Ellington, &

Grandinetti, 1998; Zhong, Frases, Wang, Casadevall, & Stark, 2008; Zujovic, Chen, & Melton, 2016). For example, Wang & Hong (2016) recently introduced multidimensional solid-state NMR spectroscopy coupled with the ^{13}C labeling of whole plants to determine the spatial arrangements of macromolecules in near-native plant cell walls (Wang & Hong, 2016). Using ^{13}C solid-state NMR, the molecular structure and dynamics changes occurring upon hydration and upon gelation for locust bean gum and konjac glucomannan were studied.

However, it should be noted that solid-state NMR normally gives poor resolution due to the full effects of anisotropic or orientation-dependent interactions, therefore, the enhancement of ^{13}C is very important. For example, as ^{13}C labeling gives the necessary sensitivity to correlate and resolve the signals of many polysaccharides and proteins, Wang and Hong (2016) labeled arabidopsis and brachypodium primary cell walls by growing the plants in liquid culture containing ^{13}C-labelled glucose in the dark. You may refer to the review paper of Paasch and Brunner (2010) for the detailed introduction of solid state NMR.

References

Agrawal, P. K. (1992). NMR spectroscopy in the structural elucidation of oligosaccharides and glycosides. *Phytochemistry, 31*(10), 3307–3330.

Bootten, T. J., Harris, P. J., Melton, L. D., & Newman, R. H. (2009). Solid-State C-13 NMR study of a composite of tobacco Xyloglucan and gluconacetobacter xylinus cellulose: Molecular interactions between the component polysaccharides. *Biomacromolecules, 10*(11), 2961–2967.

Bubb, W. A. (2003). NMR spectroscopy in the study of carbohydrates: Characterizing the structural complexity. *Concepts in Magnetic Resonance Part A, 19A*(1), 1–19.

Cui, S. (2005a). Structural analysis of polysaccharides. In S. Cui (Ed.), *Food carbohydrates: Chemistry, physical properties, and applications* (pp. 106–160). CRC Press.

Dick-Perez, M., Wang, T., Salazar, A., Zabotina, O. A., & Hong, M. (2012). Multidimensional solid-state NMR studies of the structure and dynamics of pectic polysaccharides in uniformly C-13-labeled Arabidopsis primary cell walls. *Magnetic Resonance in Chemistry, 50*(8), 539–550.

Dick-Perez, M., Zhang, Y., Hayes, J., Salazar, A., Zabotina, O. A., & Hong, M. (2011). Structure and interactions of plant cell-wall polysaccharides by two- and three-dimensional magic-angle-spinning solid-state NMR. *Biochemistry, 50*(6), 989–1000.

Guo, Q., Cui, S. W., Wang, Q., Hu, X., Wu, Y., Kang, J., et al. (2011). Structure characterization of high molecular weight heteropolysaccharide isolated from Artemisia sphaerocephala Krasch seed. *Carbohydrate Polymers, 86*(2), 742–746.

Ivanir, H., & Goldbourt, A. (2014). Solid state NMR chemical shift assignment and conformational analysis of a cellulose binding protein facilitated by optimized glycerol enrichment. *Journal of Biomolecular NMR, 59*(3), 185–197.

Ng, J. K. T., Zujovic, Z. D., Smith, B. G., Johnston, J. W., Schroeder, R., & Melton, L. D. (2014). Solid-state C-13 NMR study of the mobility of polysaccharides in the cell walls of two apple cultivars of different firmness. *Carbohydrate Research, 386*, 1–6.

Paasch, S., & Brunner, E. (2010). Trends in solid-state NMR spectroscopy and their relevance for bioanalytics. *Analytical and Bioanalytical Chemistry, 398*(6), 2351–2362.

Pizzoferrato, L., Manzi, P., Bertocchi, F., Fanelli, C., Rotilio, G., & Paci, M. (2000). Solid-state (13)C CP MAS NMR spectroscopy of mushrooms gives directly the ratio between proteins and polysaccharides. *Journal of Agriculture and Food Chemistry, 48*(11), 5484–5488.

Qian, K.-Y., Cui, S. W., Nikiforuk, J., & Goff, H. D. (2012). Structural elucidation of rhamnogalac-turonans from flaxseed hulls. *Carbohydrate Research, 362,* 47–55.

Redenti, E., Zanol, M., Ventura, P., Fronza, G., Comotti, A., Taddei, P., et al. (1999). Raman and solid state 13C-NMR investigation of the structure of the 1:1 amorphous piroxicam: beta-cyclodextrin inclusion compound. *Biospectroscopy, 5*(4), 243–251.

Spevacek, J., & Brus, J. (2008). Solid-state NMR studies of polysaccharide systems. *Macromolecular Symposia, 265,* 69–76.

Tzou, D.-L. M. (2005). A solid-state NMR application of the anomeric effect in carbohydrates: Galactosamine, glucosamine, and N-acetyl-glucosamine. *Solid State Nuclear Magnetic Resonance, 27*(4), 209–214.

Vieira, M. C., & Gil, A. M. (2005). A solid state NMR study of locust bean gum galactomannan and Konjac glucomannan gels. *Carbohydrate Polymers, 60*(4), 439–448.

Wang, T., & Hong, M. (2016). Solid-state NMR investigations of cellulose structure and interactions with matrix polysaccharides in plant primary cell walls. *Journal of Experimental Botany, 67*(2), 503–514.

Wawer, I., Wolniak, M., & Paradowska, K. (2006). Solid state NMR study of dietary fibre powders from aronia, bilberry, black currant and apple. *Solid State Nuclear Magnetic Resonance, 30*(2), 106–113.

White, P. B., Wang, T., Park, Y. B., Cosgrove, D. J., & Hong, M. (2014). Water-Polysaccharide interactions in the primary cell wall of *Arabidopsis thaliana* from polarization transfer solid-state NMR. *Journal of the American Chemical Society, 136*(29), 10399–10409.

Zhang, P., Klymachyov, A. N., Brown, S., Ellington, J. G., & Grandinetti, P. J. (1998). Solid-state 13C NMR investigations of the glycosidic linkage in alpha-alpha' trehalose. *Solid State Nuclear Magnetic Resonance, 12*(4), 221–225.

Zhong, J., Frases, S., Wang, H., Casadevall, A., & Stark, R. E. (2008). Following fungal melanin biosynthesis with solid-state NMR: Biopolyrner molecular structures and possible connections to cell-wall polysaccharides. *Biochemistry, 47*(16), 4701–4710.

Zujovic, Z., Chen, D., & Melton, L. D. (2016). Comparison of celery (*Apium graveolens* L.) collenchyma and parenchyma cell wall polysaccharides enabled by solid-state C-13 NMR. *Carbohydrate Research, 420,* 51–57.

Chapter 8
MALDI-TOF-MS for Polysaccharides Analysis

8.1 Introduction

Mass spectroscopy has been widely applied for carbohydrate analysis in terms of molecular mass, constituent monosaccharides, linkage types, sequencing of sugar residues, branching features, types of modifying groups, and the quantity (Cui, 2005b; Kailemia, Ruhaak, Lebrilla, & Amster, 2014; Madson, 2016; Mutenda & Matthiesen, 2007).

The most widely used ionization methods for oligosaccharides are matrix-assisted laser desorption/ionization (MALDI) and electrospray ionization (ESI). Compared to the traditional fast atom bombardment (FAB), they impart less energy and could better maintain the integrity of the fragment during ionization. MALDI, compared to ESI, has higher sensitivity for glycans, ionizes well even at higher mass range, and it could tolerate contaminants. The MALDI-TOF-MS method, therefore, has been extensively selected for the structural characterization of polysaccharides, although the pre-sample treatment such as controlled degradation and/or pre-methylation is required. In the current chapter, MALDI-TOF-MS in terms of principles and procedures are focused, while its polysaccharides application is discussed as well.

8.2 Principles and Procedures (Cui, 2005a)

(1) **Sample preparation**: Sample preparation involves mixing the polysaccharide molecules with matrix and formation of a "solid solution" or suspension. The polysaccharide molecules are required to be monomerly dispersed in the matrix so that they are completely isolated from each other. A homogenous "solid solution" is formed by evaporating any liquid solvent(s) used in the preparation of the solution before analysis. Here the matrix should meet the following requirements: (1) it should be soluble in solvents, compatible with analyte and

© Crown 2018

Q. Guo et al., *Methodology for Structural Analysis of Polysaccharides,*
Biobased Polymers, https://doi.org/10.1007/978-3-319-96370-9_8

vacuum stable; (2) it could absorb the laser wavelength and cause co-desorption of the analyte upon laser irradiation and promote analyte ionization. In general, aromatic acids with a chromophore are typical matrixes for ultraviolet lasers. The mid-infrared laser wavelengths are also possible for MALDI.

(2) **Matrix excitation:** UV or IR laser light is used in this step to irradiate a portion (usually about 100 μm in diameter) of the sample/matrix mixture. Some of the laser energy incident on the solid solution is absorbed by the matrix, causing rapid vibrational excitation and ionization.

(3) **Analyte ionization and detection:** some of the charges in the photo-excited matrix are passed to the analyte molecules and forms clusters of single ana- lyte molecules surrounded by neutral and excited matrix clusters. The matrix molecules are evaporated to give an excited analyte molecule, hence, lead to the formation of the typical $[M + X]^+$ type of ions (where $X = H$, Li, Na, K, etc.) and subsequently detected by a mass detector. Negative ions can also be formed from reactions involving deprotonation of the analyte, which lead to the formation of $[M - H]^-$ molecular ions. Either positive or negative ions will be separated and identified by the TOF-MS detector.

8.3   Applications

The use of MALDI-TOF-MS for xyloglucan analysis has been previously reported by Ding et al. (2016). Polysaccharides extracted from flaxseed kernel were firstly converted into small molecular fragments using both enzymatic hydroly- sis and 0.1M TFA hydrolysis separately and then analysed by MALDI-TOF-MS analysis (Fig. 8.1). The matrix chosen for MALDI-TOF-MS analysis was 2,5- Dihydroxybenzoic (DHB) acid, which was dissolved in 1 mL 20% (v/v) ethanol solution. Samples were re-suspended in 1 mL matrix solution and dried on a stainless- steel plate MALDI prior to analysis. The sample/matrix mixtures were allowed to air-dry and were analysed in a Bruker Reflex III equipped with a 337 nm nitrogen laser (Bruker, Germany). Spectra were collected from an average of at least 100 shots (Ding et al., 2015).

It was found that enzymatic hydrolysis coupled with MALDI-TOF analysis (Table 8.1; Fig. 8.1) was proved to be a better method of structural analysis for polysaccharides. Fragments of XXXG, XXLG, XLLG substitution patterns were detected with a relatively high percentage, while other fragments including GXXXG, GXXG, XXFG etc. are all detected and listed in Table 8.1. Here G represent β-D-GlcP backbone or reducing end; X represent T-α-D-XylP-(1→6)-linked with G; L represent T-β-D-GalP-(1→2)-α-D-XylP-(1→6)-linked with G; F represent T-α-L-FucP-(1→2)-β-D-GalP-(1→2)-α-D-XylP-(1→6)-linked with G; S represent T-α-L-Araf-(1→2)-α-D-XylP-(1→6)-linked with G.

Table 8.1 MALDI-TOF-MS analysis and deduced subunit structures of xyloglucans from flaxseed kernel cell wall

KPI-EPF-G$^a_{24h}$			KPI-EPF-H$^b_{1.5h}$		
m/z	Ion	Structure^c	m/z	Ion	Structure^c
791.5	M + Na$^+$	XXG	791.5	M + Na$^+$	XXG
–	–	–	804.8	M + Na$^+$	FG
807.7	M + K$^+$ (weak)	XXG	807.7	M + K$^+$ (weak)	XXG
–	–	–	820.6*	M + Na$^+$	GGXG*
927.9*	M + Na$^+$ (weak)	XSG*	–	–	–
–	–	–	937.1	M + Na$^+$	XF
953.7	M + Na$^+$	XXGG/GXXG	953.7	M + Na$^+$	XXGG/GXXG
969.3	M + K$^+$ (weak)	XXGG/GXXG	969.3	M + K$^+$	XXGG/GXXG
–	–	–	983.0*	M + Na$^+$	GGLG*
1085.9	M + Na$^+$	XXXG	1085.9	M + Na$^+$	XXXG
–	–	–	1099.2	M + Na$^+$	XFG
1102.8	M + K$^+$ (weak)	XXXG	1102.8	M + K$^+$	XXXG
1116.1	M + Na$^+$	GXLG/GXXGG	1116.1	M + Na$^+$	GXLG/GXXGG
1133.6	M + K$^+$ (weak)	GXLG/GXXGG	1133.6	M + K$^+$	GXLG/GXXGG
–	–	–	1145.4*	M + Na$^+$	GGLGG*
1232.5	M + Na$^+$	XXF	1232.5	M + Na$^+$	XXF
1248.3	M + Na$^+$ and/or M + K$^+$	XXLG/GXXXG, and/or XXF	1248.3	M + Na$^+$ and/or M + K$^+$	XXLG/GXXXG, and/or XXF
–	–	–	1261.9	M + Na$^+$	GXFG/LFG
1278.2	M + Na$^+$ (weak)	GLLG	1278.2	M + Na$^+$	GLLG
–	–	–	1292.0*	M + Na$^+$	GGFGG*
–	–	–	1307.7*	M + Na$^+$	GGGLGG*
1395.0	M + Na$^+$	XXFG	1395.0	M + Na$^+$	XXFG
1410.9	M + Na$^+$ and/or M + K$^+$	XLLG/GXXLG and/or XXFG	1410.9	M + Na$^+$ and/or M + K$^+$	XLLG/GXXLG and/or XXFG
–	–	–	1424.7*	M + Na$^+$	GLFG/GXFGG*
1441.6	M + Na$^+$ (weak)	GLLGG	1441.6	M + Na$^+$	GLLGG
–	–	–	1454.4*	M + Na$^+$	GGGFGG*
–	–	–	1470.3*	M + K$^+$	GGGFGG*
1547.2*	M + Na$^+$ (weak)	GLXSG*	–	–	–
1558.9	M + Na$^+$ (weak)	XLFG	1558.9	M + Na$^+$	XLFG
1574.1	M + Na$^+$ and/or M + K$^+$	GXLLG and/or XLFG	1574.1	M + Na$^+$ and/or M + K$^+$	GXLLG and/or XLFG
1586.0	M + Na$^+$ (weak)	GLFGG	1586.0	M + Na$^+$	GLFGG
1603.2	M + Na$^+$ (weak)	GGLLGG	1603.2	M + Na$^+$	GGLLGG
–	–	–	1617.7*	M + Na$^+$	GGGFGGG*
1707.5*	M + Na$^+$ (weak)	XXLLG*	1707.5*	M + Na$^+$ (weak)	XXLLG*
–	–	–	1719.9*	M + Na$^+$	GXLFG*
–	–	–	1736.1*	M + K$^+$ (weak)	GXLFG*

Adapted from Ding et al. (2016)

* The m/z value might be contributed by other sugar units from contaminants and could cause the false positive results

^{a} KPI-EPF-G$_{24h}$: endo-1,4-β-D-glucanase hydrolysis of KPI-EPF for 24 h

^{b} KPI-EPF-H$_{1.5h}$: 0.1 M TFA hydrolysis of KPI-EPF for 1.5 h

^{c} G: β-D-Glc*p* backbone or reducing end; X: T-α-D-Xyl*p*-(1→6)-linked with G; L: T-β-D-Gal*p*-(1→2)-α-D-Xyl*p*-(1→6)-linked with G; F: T-α-L-Fuc*p*-(1→2)-β-D-Gal*p*-(1→2)-α-D-Xyl*p*-(1→6)-linked with G; S: T-α-L-Ara*f*-(1→2)-α-D-Xyl*p*-(1→6)-linked with G

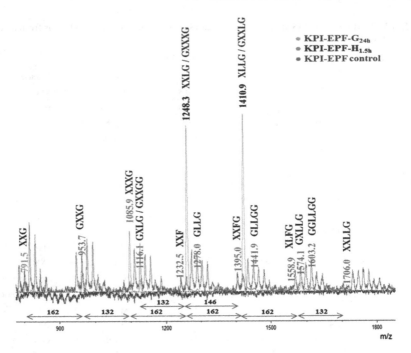

Fig. 8.1 MALDI-TOF-MS profile of KPI-EPF (KPI-EPF-G refers to endo-1,4-β-D-glucanase hydrolysis of KPI-EPF for 24 h as the spectrum in orange; KPI-EPF-$H_{1.5h}$ refers to 0.1 M TFA hydrolysis of KPI-EPF for 1.5 h as the spectrum in green; KPI-EPF control refers to no treatment of KPI-EPF as the spectrum in blue) [M + Na/K]$^{+}$ was detected. G: β-D-Glc$p$ backbone or reducing end; X: T-α-D-Xyl$p$-(1→6)-linked with G; L: T-β-D-Gal$p$-(1→2)-α-D-Xyl$p$-(1→6)-linked with G; F: T-α-L-Fuc$p$-(1→2)-β-D-Gal$p$-(1→2)-α-D-Xyl$p$-(1→6)-linked with G; S: T-α-L-Ara$f$-(1→2)-α-D-Xyl$p$-(1→6)-linked with G. This figure is adapted from Ding et al. (2016)

References

Cui, S. (2005a). Structural analysis of polysaccharides. In S. Cui (Ed.), *Food carbohydrates: Chemistry, physical properties, and applications* (pp. 106–160). CRC Press.

Cui, S. W. (2005b). *Structural analysis of polysaccharides. Food carbohydrates: Chemistry, physical properties, and applications* (pp. 105–157). Boca Raton, Florida: Taylor and Francis.

Ding, H. H., Cui, S. W., Goff, H. D., Chen, J., Guo, Q., & Wang, Q. (2016). Xyloglucans from flaxseed kernel cell wall: Structural and conformational characterization. *Carbohydrate Polymers, 151*, 538–545.

Ding, H. H., Cui, S. W., Goff, H. D., Chen, J., Wang, Q., & Han, N. F. (2015). Arabinan-rich rhamnogalacturonan-I from flaxseed kernel cell wall. *Food Hydrocolloids, 47*, 158–167.

Kailemia, M. J., Ruhaak, L. R., Lebrilla, C. B., & Amster, I. J. (2014). Oligosaccharide analysis by mass spectrometry: A review of recent developments. *Analytical Chemistry, 86*(1), 196–212.

Madson, M. A. (2016). *Mass Spectral Analysis of Carbohydrates. Mass Spectrometry* (pp. 1–78). Boston: Elsevier.

Mutenda, K., & Matthiesen, R. (2007). Analysis of carbohydrates by mass spectrometry. In R. Matthiesen (Ed.), *Methods in molecular biology* (Vol. 367). Totowa, NJ: Humana Press Inc.

Chapter 9
Fourier Transform Infrared Spectroscopy (FTIR) for Carbohydrate Analysis

As FTIR spectrometer can simultaneously collect high-resolution data over a wide spectral range in a short time, it has been extensively used for carbohydrate analysis (Guo et al., 2011, 2015; Kang et al., 2011; Singthong, Ningsanond, Cui, & Douglas Goff, 2005). FTIR spectra in the wavenumber between 950 and 1200 cm^{-1} are considered as the 'fingerprint' region for carbohydrates, which allows the identification of major chemical groups in polysaccharides as the position and intensity of the bands are specific for every polysaccharide. The other typical FT-IR spectra for polysaccharides are listed in Table 9.1

However, the biggest advantage of FTIR for carbohydrate analysis is its ability to identify and quantify the functional groups. For example, pectin have the methyl ester character and the degree of esterification can be quickly estimated using FTIR spectroscopy (Guo et al., 2015; Singthong et al., 2005). Peaks at 1742 and 1604 cm^{-1} are derived from ester carbonyl groups and carboxylate ion stretching band, respectively (Fig. 9.1), the DE value linear correlates with the value of $A_{1742}/(A_{1742} + A_{1604})$ (here A refer to peak area) according to Singthong, Cui, Ningsanond, and Douglas Goff (2004), by which the DE value of GSP (pectic polysaccharides from American ginseng) was calculated to be 38% (Guo et al., 2015) as shown in Fig. 9.1.

Table 9.1 Typical FTIR spectrum for functional groups of polysaccharides

Wavenumber (cm^{-1})	Functional groups/vibration mode	Intensity
3500–3000	O–H stretching	Broad, strong
3000–2800	C–H stretching, symmetric, asymmetric	Sharp, occasionally double overlapping with O–H
1630–1600	COO-asymmetric stretching	Strong
1400	COO-symmetric stretching	Weak
1380	C–H bending	Weak
1280–900	Finger-print of carbohydrates	Strong

© Crown 2018

Q. Guo et al., *Methodology for Structural Analysis of Polysaccharides,*
Biobased Polymers, https://doi.org/10.1007/978-3-319-96370-9_9

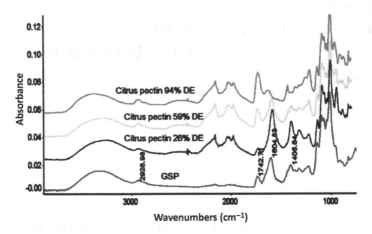

Fig. 9.1 FTIR spectra of from commercial citrus pectin of known DE and pectin from American ginseng. Adapted from Guo et al. (2015)

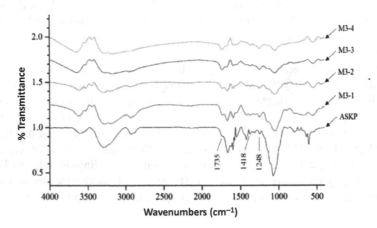

Fig. 9.2 FTIR spectra of ASKP and acetylated ASKPs. M3-1, M3-2, M3-3 and M3-4 were obtained by the method of acetic anhydride-pyridine system with formamide as the solvent. Degrees of substitution were calculated as 0.04, 0.13, 0.31 and 0.42, respectively. Adapted from Li et al. (2016)

FTIR spectrum also can be applied for monitoring the structural modification of polysaccharides. As shown in Fig. 9.2, the effects on structural characteristics of acetylated ASKPs through reaction with acetic anhydride were analyzed by FTIR spectroscopy (Li, Hu, Li, & Ma, 2016). Three absorption bands at 1735 cm^{-1} (C=O ester stretching vibration), 1418 cm^{-1} (C–H bending vibration in –OAc and 1248 cm^{-1} (C–O ester stretching vibration) were used to evaluate the degree of acetylation (DS). DS affected by the acetic anhydride ratio was investigated by the relative intensity of the absorption bands at 1735, 1418, and 1248 cm^{-1}. The relative intensity of the three bands increased with the increase of the acetic anhydride ratio, indicating the improvement of DS.

References

Guo, Q. B., Cui, S. W., Kang, J., Ding, H. H., Wang, Q., & Wang, C. (2015). Non-starch polysaccharides from American ginseng: Physicochemical investigation and structural characterization. *Food Hydrocolloids, 44,* 320–327.

Guo, Q., Cui, S. W., Wang, Q., Hu, X., Guo, Q., Kang, J., & Yada, R. (2011). Extraction, fractionation and physicochemical characterization of watersoluble polysaccharides from Artemisia sphaerocephala Krasch seed. *Carbohydrate Polymers, 86*(2), 831–836.

Kang, J., Cui, S. W., Chen, J., Phillips, G. O., Wu, Y., & Wang, Q. (2011). New studies on gum ghatti (*Anogeissus latifolia*) part I. Fractionation, chemical and physical characterization of the gum. *Food Hydrocolloids, 25*(8), 1984–1990.

Li, J., Hu, X., Li, X., & Ma, Z. (2016). Effects of acetylation on the emulsifying properties of Artemisia sphaerocephala Krasch. polysaccharide. *Carbohydrate Polymers, 144,* 531–540.

Singthong, J., Cui, S. W., Ningsanond, S., & Douglas Goff, H. (2004). Structural characterization, degree of esterification and some gelling properties of Krueo Ma Noy (*Cissampelos pareira*) pectin. *Carbohydrate Polymers, 58*(4), 391–400.

Singthong, J., Ningsanond, S., Cui, S. W., & Douglas Goff, H. (2005). Extraction and physicochemical characterization of Krueo Ma Noy pectin. *Food Hydrocolloids, 19*(5), 793–801.

Chapter 10
Detailed Experimental Procedures

10.1 Total Sugar Analysis (Phenol-Sulfuric Acid Assay)

10.1.1 Sample Preparation

Weigh 10 mg polysaccharide sample, adding 1 mL 12 M sulphate acid and keep stirring at room temperature for 30 min until sample dissolved well. Dilute the solution accordingly, normally 10 times and waiting for test.

10.1.2 Testing Method

Phenol (50 μL, 80%) is added to a testing tube with 2.0 mL clear sample solution (the sample should be well dissolved) followed by adding 3.0 mL concentrated sulphuric acid. After thoroughly mixing on a vortex, the solution is kept at room temperature for a sufficient time (usually more than 20 min) to allow for the color development. The standing time should be kept the same for all the samples to ensure the reproducibility. The absorbance is determined by a spectrophotometer, at 490 or 480 nm depending on the types of sugar present. A series of different concentration of standard (such as 50, 100, 150 and 200 μg/mL) are usually used to construct a calibration curve of absorbance versus concentration. Glucose is usually adopted as the standard for hexosyl sugars, and the data obtained should be reported as glucose equivalence. However, a standard mixture can be used to obtain more accurate results if the estimated monosaccharide ratio in measured samples is known (DuBois, Gilles, Hamilton, Rebers, & Smith, 1956).

© Crown 2018
Q. Guo et al., *Methodology for Structural Analysis of Polysaccharides*,
Biobased Polymers, https://doi.org/10.1007/978-3-319-96370-9_10

10.2 Uronic Acid Analysis

10.2.1 Sample Preparation

Samples containing uronic acid are firstly dissolved in water and high temperature (50 °C) is usually used. The sample solution is diluted so that the concentration of uronic acids is within the linear region of the standards curve (usually glucuronic acid with the concentration of 10, 20, 30, 40 and 50 μg/mL, respectively).

10.2.2 Testing Method

Sulfuric acid (0.0125 M) containing sodium tetraborate (prepared by adding 0.25 g sodium tetraborate into 100 mL of concentrated sulfuric acid) is thoroughly mixed with the sample solution in a test tube and placed in an oil bath at 100 °C for 5 min. After placing the test tube in ice water bath for rapid cooling, m-hydroxydiphenyl (0.15%, prepared by adding 0.15 g 3-phenylphenol and 0.5 g NaOH into 100 mL distilled water) is added and completely mixed. 20 min are usually required to allow the color to develop. The absorbance of the measured sample is read at 520 nm. A sample blank (containing a sample solvent without the m-hydroxydiphenyl) should be prepared at the same time as reference samples (Blumenkrantz & Asboe-Hansen, 1973).

10.3 Column Separation for Polysaccharides from Flaxseed Hulls

The detailed procedures are as follows (Qian, Cui, Wu, & Goff, 2012):

(1) Sample Preparation

SFG (0.35 g) was firstly dissolved in 1 L of buffer 1 (20 mM Tris/HCl, pH 8) at 80 °C for 1 h and then cooling down to room temperature.

(2) Loading and Collection of Neutral Sugar

SFG solution was loaded onto the pre-equilibrated column (XK 50Column, GE Healthcare, packed with 900 mL of Q-Sepharose fast flow as the matrix) at a flow rate of 10 mL/min for 100 min and flushed continuously with 2 L of buffer 1 for 200 min. The eluate (denoted as NFG) was collected between 25 and 150 min after the sample loading.

(3) Collection of Acidic Sugar

When w90 wt% of NFG was eluted out, the column was successively washed with 1 L of buffer 2 (20 mM Tris/HCl + 1 M NaCl, pH 8) at the same flow rate (10 mL/min).

The eluate (denoted as AFG) was collected between 55 and 110 min after buffer 2 was loaded, during which period most of the acidic fraction (AFG) was eluted out.

(4)   Column Re-equilibration

The column was sequentially flushed with 1 L of each of the following three washing solutions: 2 M NaCl, 1 M NaOH and re-equilibrated with 2–3 L of buffer 1 for next use.

10.4   Methylation Analysis

If the sample contains a high percentage of uronic acids, it will create difficulties for methylation analysis. Therefore, uronic acid was first reduced into neutral sugars prior to methylation analysis. The related steps are described as follows.

10.4.1   *Reduction of Uronic Acids*

(1)   Sample Preparation and Activation

The acidic polysaccharide (5 mg) was dissolved in deuterium oxide D_2O (2 mL). To the solution, 50 mg of 1-cyclohexyl-3-(2-morpholinoethyl)-carbodimidemethyl-p-toluenesulfonate is added while using 0.1 mol/L HCl in D2O to keep the pH at 4.75. The solution was left for 1 h under stirring.

(2)   Reduction

5 mL of sodium borodeuteride (160 mg/mL) is added into solution dropwise over a period of 1 h, and the reaction mixture pH was maintained at 7.0, using 2.0 mol/L HCl in D_2O during the reduction reaction. The reaction continues with constant stirring for 0.5 h at pH 7.0 after the addition of sodium borodeuteride, then the solution pH was brought back to 4.0. The reduced polysaccharide is separated from salts by dialysis against distilled water overnight at 25 °C (3500 Da molecular weight cut off), then lyophilized.

(3)   Sample Purification and Preparation for the Methylation Analysis

The polysaccharide is re-dissolved in 1 mL distilled water and 0.5 mL 10% acetic acid in methanol is added. The mixture is dried under a stream of nitrogen to remove the boric acid. Another 1 mL of 10% acetic acid in methanol is added to the residue and evaporated using nitrogen. This process is repeated 3–4 times to ensure that most of the boric acid is removed. Finally, a few drops of methanol are added and the solution was evaporated (two times) to remove any boric acid remaining.

10.4.2 Methylation Analysis

(1) Polymer Dissolution and Methylation

Dissolve 2–3 mg dried polysaccharides (vacuum drying at 80 °C for 5 h) into 0.5 mL DMSO, heating and sonication are sometimes required to aid the solubilization. Add dried and very fine sodium hydroxide (20 mg) and 0.3 mL methyl iodide to the reaction vial; keep stirring to carry out methylation reaction, normally takes 2.5 h. Transfer the reactant solution to a 25–50 mL flat glass vial (wash the reaction vial using each 1–2 mL methylene chloride for three times). Adding water 3–5 mL into the glass vial followed by agitation and waiting for the phase separation, removing the top aqueous layer, repeating this process for three times. The organic layer is filtered through a sodium sulphate column into 5 mL reaction vial, followed by nitrogen drying.

(2) Hydrolysis and Reduction

Adding 4 M trifluoroacetic acid (TFA) to a sealed reaction vial at 100 °C for 6 h to convert the methylated polysaccharides to monomers. Removing the TFA through a continuous stream of nitrogen, add 0.3 mL deionized water, one drop of 1% ammonium hydroxide and approximate 3 mg sodium borodeuteride/D_2O solution to conduct the reduction analysis for 12 h.

(3) Acetylation

Adding acetic acid dropwise to stop the reduction reaction, add 0.5 mL 5% acetic acid in methanol, vortexed and evaporated until dry by continuous nitrogen stream. Repeating this step with 100% methanol and dried with nitrogen for several times. 0.5 mL acetic anhydride is added and the mixture heated at 100 °C for 2 h. The extra acetic anhydride after reaction can be destroyed by a few drops of ethanol; the reaction solution once again was dried using nitrogen. Methylene chloride (0.5 mL × 2) was added to the vial to dissolve the partially methylated alditol acetates (PMAA), and wait for GC-MS analysis.

(4) GC-MS Analysis

The capillary gas chromatography for separation and quantification of the partially methylated alditol acetates (PMAA) has been previously described (Sweet, Albersheim, & Shapiro, 1975). The relative retention times and elution order of some of the PMAAs will depend upon the coating used in the GC column. One detailed experimental procedure for GC-MS of PMAA is listed as follows (Guo et al., 2011): aliquots of the resultant PMAAs are injected into a GC-MS system (ThermoQuest Finnigan, San Diego, CA) fitted with a SP-2330 (Supelco, Bellefonte, Pa) column (30 mL × 0.25 mm, 0.2 mm film thickness, 160–210 °C at 2 °C/min, then 210–240 °C at 5 °C/min) and an ion trap MS detector.

10.5 Controlled Degradation

10.5.1 Periodate Oxidation and Smith Degradation (Ai et al., 2016)

(1) Periodate Oxidation

Polysaccharides (30 mg) is firstly dissolved in 30 mL 0.03 M NaIO$_4$ solution and kept in dark at 4 °C. The reaction was monitored by tracking the OD value at 230 nm every 12 h until stable; 10 mL sample was taken and mixed with 1 mL ethylene glycol for about 10 min to reduce NaIO$_4$ residue in solution. Formic acid is determined by direct titration with standard NaOH (0.1 M).

(2) Reduction and Mild Acid Hydrolysis

After periodate oxidation, 20 mL solution was taken and mixed with 2 mL ethylene glycol to react with the excess periodate. The solution was kept for 30 min at 25 °C, then dialyzed against distilled water to remove inorganic salts. 100 mg of sodium borohydride is added to the dialyzed solution (500 mL) to reduce the aldehyde group. The reduction reaction was continued for about 20 h at 25 °C and the excess borohydride was destroyed by dropwise addition of hydrochloric acid (1 M). The neutral solution was dialyzed against distilled water and lyophilized, and then 1 mg dried sample is dissolved into 2 mL 2 M TFA and hydrolysed at 121 °C for 1 h.

(3) Detection

The dried hydrolysates are prepared by evaporation under nitrogen, and then mixed with 10 mg hydroxylamine hydrochloride and 0.5 mL anhydrous pyridine under 90 °C for 30 min. The acetylation is carried out by reacting with anhydrous acetyl anhydride at 90 °C for 30 min. GC analysis of the acetylated samples is carried out on a OV-1701 column—30 m × 0.35 mm; H$_2$ 0.6 kg/cm^2 and air 0.5 kg/cm^2; 150–190 °C at 10 °C/min for 1 min, then 190–240 °C at 3 °C/min for 20 min; an FID detector.

10.5.2 Partial Acid Hydrolysis

Sample (40 mg) is hydrolyzed with 0.1 M trifluoroacetic acid (TFA) (10 mL) at 100 °C for different time, e.g. 0.5, 1.5, 2.5 and 3.5 h, respectively. After hydrolysis, each solution is allowed to cool down to room temperature and is added 3 times of ethanol (v/v) to induce precipitation. The precipitates are separated from the supernatant by centrifugation. The polysaccharides in the precipitate/supernatant after the 0.5, 1.5, 2.5 and 3.5 h TFA treatments are freeze-dried and kept for molecular weight and structure analysis (Guo et al., 2012).

10.5.3 Trifluoromethanesulfonic Acid (TFMS) Hydrolysis

The detailed procedures are as follows (Sojar & Bahl, 1987): The glycoprotein sample is placed in an ethanol/dry ice bath and left to cool for 20 s. 5 ml TFMS and toluene solution with a ratio of 9:1 (v/v) is added slowly to the sample solution. The sample vials are kept at −20 °C for 10 min and shaken briefly to aid melting of the contents every five minutes. Then the mixture is stored in a fridge for 4 h. The excess of TFMS is neutralized by 2% (w/v) of Tris base solution. The released peptides/oligosaccharides are then collected and purified for structural analysis

10.6 Sample Preparation and Experimental Procedure for NMR Analysis (Guo et al., 2015)

40 mg polysaccharide sample is dissolved in 4 mL D_2O at 25 °C under stirring for 2 h, and then freeze-dried. This procedure is repeated three times. Samples are then dissolved in 3 mL D_2O and ready for testing. High-resolution ^1H and ^{13}C NMR spectra were recorded at 500.13 and 125.78 MHz, respectively, on a Bruker ARX500 NMR spectrometer operating at 30 °C. A 5 mm inverse geometry $^1H/^{13}C/^{15}N$ probe is used. Chemical shifts are reported relative to trimethylsilyl propionate (TSP) in D_2O for ^1H (0.0 ppm, external standard) and 1,4-Dioxane in D_2O for ^{13}C (66.5 ppm, external standard). Homonuclear $^1H/^1H$ correlation spectroscopy (COSY, TOCSY) and heteronuclear $^1H/^{13}C$ correlation experiments (HMQC, HMBC) are running using the standard Bruker pulse sequence.

References

Ai, L. Z., Guo, Q. B., Ding, H. H., Guo, B. H., Chen, W., & Cui, S. W. (2016). Structure characterization of exopolysaccharides from Lactobacillus casei LC2W from skim milk. *Food Hydrocolloids*, *56*, 134–143.

Blumenkrantz, N., & Asboe-Hansen, G. (1973). New method for quantitative determination of uronic acids. *Analytical Biochemistry, 54*(2), 484–489.

DuBois, M., Gilles, K. A., Hamilton, J. K., Rebers, P. A., & Smith, F. (1956). Colorimetric method for determination of sugars and related substances. *Analytical Chemistry, 28*(3), 350–356.

Guo, Q. B., Cui, S. W., Kang, J., Ding, H. H., Wang, Q., & Wang, C. (2015). Non-starch polysaccharides from American ginseng: Physicochemical investigation and structural characterization. *Food Hydrocolloids, 44*, 320–327.

Guo, Q., Cui, S. W., Wang, Q., Hu, X., Kang, J., & Yada, R. Y. (2012). Structural characterization of a low-molecular-weight heteropolysaccharide (glucomannan) isolated from *Artemisia sphaerocephala* Krasch. *Carbohydrate Research, 350*, 31–39.

Guo, Q., Cui, S. W., Wang, Q., Hu, X., Wu, Y., Kang, J., & Yada, R. (2011). Structure characterization of high molecular weight heteropolysaccharide isolated from *Artemisia sphaerocephala* Krasch. seed. *Carbohydrate Polymers, 86*(2), 742–746.

Qian, K. Y., Cui, S. W., Wu, Y., & Goff, H. D. (2012). Flaxseed gum from flaxseed hulls: Extraction, fractionation, and characterization. *Food Hydrocolloids, 28*(2), 275–283.

Sojar, H. T., & Bahl, O. P. (1987). A chemical method for the deglycosylation of proteins. *Archives of Biochemistry and Biophysics, 259*(1), 52–57.

Sweet, D. P., Albersheim, P., & Shapiro, R. H. (1975). Partially ethylated alditol acetates as derivatives for elucidation of the glycosyl linkage-composition of polysaccharides. *Carbohydrate Research, 40*(2), 199–216.

Chapter 11
Summary

This book provides a comprehensive review of the state of the art on the structure analysis of complex polysaccharides. Unlike protein and nucleic acid, the structure investigation of polysaccharides can only be termed as structural characterization rather than molecular sequencing due to the complexity and irregularity of the molecular chain. The obtained molecular structure can only be referred as "the proposed structure".

Several techniques for the polysaccharides structural characterization have been described in detail. It should be noted that the addressed techniques are not isolated. Instead, they are carried out simultaneously in supporting/confirming each other to elucidate the structural information. For example, the extraction methods, e.g. solvent selected, not only impact the purity, but also highly affect the structural features of obtained polysaccharides. Therefore, the selection of extraction methods should always meet their specific requirements of the application, and can be extremely useful in the purification/fractionation procedure of the polysaccharides. Similarly, structural degradation is always followed by molecular weight determination and/or MALDI-TOF-MS to monitor the molecular chain change. Methylation analysis works well with 1D and 2D NMR, as the structural features provided by both methods are always complementary to each other. Likewise, polysaccharides purification and fractionation need to be evaluated by molecular weight distribution test (HPSEC), and monosaccharides composition analysis etc.

The development of the analytical techniques has led to a rapid progress for the structural characterization of polysaccharides in the past decade. However, the establishment of the structural and functional relationships is still challenging as many of the techniques and equipments are still unique. Various polysaccharides demonstrated similar bioactivities, for example, α-glucan from Cordyceps, β-glucan from Ganoderma, (acetyl) glucomannan from Dendrobium Officinale and pectic polysaccharides from American Ginseng all were reported for their immune-

© Crown 2018

Q. Guo et al., *Methodology for Structural Analysis of Polysaccharides,*
Biobased Polymers, https://doi.org/10.1007/978-3-319-96370-9_11

modulating effects, although the structural features were totally different from each other (Guo & Cui, 2014). Therefore, there is still a long way to bridge the gap between the molecular structure of polysaccharides and their bioactivities.

Reference

Guo, Q., & Cui, S. (2014). Polysaccharides from *Dendrobium officinal*, *Cordyceps sinensis* and *Ganoderma*: Structure and bioactivities. *Gums and Stabilizers for the Food Industry, 17,* 303–317.

Printed in the United States
by Bookmasters

Printed in the United States
By Bookmasters